TIK-SCHRIFTENREIHE NR. 45

Michael H. Eisenring

Communication Channel Synthesis for Heterogeneous Embedded Systems

vdf Hochschulverlag AG an der ETH Zürich

A dissertation submitted to the
Swiss Federal Institute of Technology Zurich
for the degree of Doctor of Technical Sciences

Diss. ETH No. 14640

Prof. Dr. Lothar Thiele, examiner
Prof. Dr. Wolfgang Fichtner, co-examiner

Examination date: May 6, 2002

ISBN 3 7281 2867 8

www.vdf.ethz.ch
verlag@vdf.ethz.ch

Abstract

Embedded systems have become part of our daily live. They are the backbone for applications such as dishwashers, set top boxes, and patient monitoring systems. During their development phase the heterogeneity and connectivity of the system components is a big challenge as their interplay provides the system behavior. The communication between interacting parts is a performance critical issue and affects features such as execution speed, power consumption, and costs. Usually, several potential implementation solutions have to be elaborated that distinguish in features such as latency, cost, and especially by a different communication infrastructure. Nevertheless, only few methodologies have emerged that support an efficient assessment and implementation of alternative implementations considering the communication features of the system components. The following issues are still open:

- There exists a lot of models to describe the implementation of an embedded system. Nevertheless, the heterogeneity of the system components prevented a consistent model (for processors, reconfigurable hardware devices, and memories) that considers their communication features as well.

- During development a behavior description is refined and mapped onto an implementation description. There exists no methodology that provides the refinement by considering the communication as well as the reconfiguration features of system components.

The subject of this research work is on models, methodologies, and implementation techniques to efficiently establish a communication infrastructure within an heterogeneous embedded system. The major contributions are:

- A set of dataflow oriented specification models to capture and describe synthesis problems for heterogeneous embedded systems. Basically, they comprise (i) an implementation independent behavior specification, (ii) a target description, as well as (iii) a mapping between behavior and target. The models base on point-to-point communication channels.

- One of these models enables the specification of dynamic reconfigured systems by the use of a hierarchical reconfiguration structure.
- Based on the specification models refinement methodologies have been proposed that consider the communication features of the system components.
- For the system components an object-oriented approach has been proposed that enables an automatic generation of interface circuitry and device drivers to establish a communication infrastructure. It is the base for (i) simple component modeling, (ii) reuse of existing generation methodologies, and (iii) simple retargeting.
- A taxonomy of communication types has been defined that summarizes the arising types of communication within an embedded system depending on the mapping.

Kurzfassung

Eingebettete Systeme sind Teil unseres täglichen Lebens geworden. Sie bilden das Rückgrat für Anwendungen wie Waschmaschinen, Digitalempfänger und Patientenüberwachungssysteme. Während ihrer Entwicklungsphase sind die Heterogenität und Verbindbarkeit der Systemkomponenten eine grosse Herausforderung, da ihr Zusammenspiel das Systemverhalten bestimmt. Die Kommunikation zwischen interagierenden Teilen hat einen starken Einfluss auf die Leistungsfähigkeit und beeinflusst Eigenschaften wie Ausführungsgeschwindigkeit, Leistungsverbrauch und Kosten.

Oft werden mehrere potenzielle Lösungen untersucht. Diese unterscheiden sich durch Eigenschaften wie Latenz, Kosten und im speziellen durch die Kommunikationsinfrastruktur. Trotzdem haben sich nur wenige Methoden durchgesetzt, welche eine effiziente Bewertung und Implementation von alternativen Lösungen erlauben und gleichzeitig die Kommunikationseigenschaften der Systemkomponenten berücksichtigen.

Die folgenden Fragestellungen sind noch immer unbeantwortet:

- Es gibt viele Modelle, um die Implementation von eingebetteten Systemen zu beschreiben. Jedoch verhinderte die Heterogenität der Systemkomponenten ein einheitliches Model (für Prozessoren, rekonfigurierbare Hardwarebausteine und Speicher), welches auch ihre Kommunikationseigenschaften berücksichtigt.

- Während der Entwicklungsphase wird eine Verhaltensbeschreibung verfeinert und auf eine Implementationsbeschreibung abgebildet. Es gibt keine Methodik, welche die Kommunikations- und Rekonfigurationseigenschaften der Systemkomponenten bei der Verfeinerung berücksichtigt.

Die vorliegende Forschungsarbeit konzentriert sich auf Modelle, Methodiken und Implementationstechniken, um effizient eine Kommunikationsinfrastruktur für eingebettete Systeme erstellen zu können. Die wesentlichen Beiträge zu diesem Forschungsgebiet sind:

- Datenflussorientierte Spezifikationsmodelle zum Erfassen und Beschreiben von Syntheseproblemen für heterogene eingebettete Systeme. Grundsätzlich umfassen sie (i) eine implementationsunabhängige Verhaltensbeschreibung, (ii) eine Beschreibung der Zielarchitektur, sowie (iii) eine Abbildung zwischen Verhaltensbeschreibung und Zielarchitektur. Die Modelle basieren auf Punkt-zu-Punkt Kommunikationskanälen.

- Eines dieser Modelle ermöglicht die Spezifikation von dynamisch rekonfigurierbaren Systemen durch die Anwendung einer hierarchischen Rekonfigurationsstruktur.

- Basierend auf den eingeführten Spezifikationsmodellen wurden Verfeinerungsmethodiken vorgeschlagen, welche die Kommunikationseigenschaften der Systemkomponenten berücksichtigen.

- Für die Systemkomponenten wurde ein objekt-orientierter Ansatz vorgestellt, welcher eine automatische Generierung von Interfaceschaltungen und Komponententreibern für die Kommunikationsinfrastruktur ermöglicht. Er ist die Basis für (i) einfache Modellierung von Systemkomponenten, (ii) Wiederverwendung von existierenden Generationsmethoden und (iii) einfache Änderung der Zielarchitektur.

- Abhängig von der Abbildung zwischen Verhaltensbeschreibung und Zielarchitektur entstehen verschiedene Kommunikationstypen. Diese wurden in einer ”Taxonomie der Kommunikationstypen” zusammengestellt.

I would like to thank

- Prof. Dr. Lothar Thiele for advising my research work and providing an interesting research environment,
- Prof. Dr. Wolfgang Fichtner for his support as a co-examiner of my thesis,
- my colleague Dr. Marco Platzner for the fruitful discussions and valuable inputs to my work, his cooperation in several papers, and his careful proof reading of this thesis,
- my family for their support during the long time I was writing on this monograph.

To my wife

Manuela

and our children

Fabienne and **Florian**

Contents

Chapter 1

Introduction

Since 1971 when INTEL introduced the first general-purpose processor 4004 in the market, embedded systems have become part of our daily life. They are the backbone for a wide range of application domains; from simple home appliances such as dishwashers, to PDAs (personal digital assistants), portable phones, and up to sophisticated patient monitoring systems in hospitals. Design, implementation as well as commercial exploitation of these systems is influenced by various factors. Technical issues such as correct functionality and power efficiency, as well as non-technical factors such as time-to-market and price have to be considered to differentiate from competitors. A big challenge during their development phase is the heterogeneity and connectivity of the system components as their interplay provides the system behavior. Nevertheless, reuse in the sense of system platforms is rather unusual as embedded systems are dedicated to a certain application.

This submitted thesis is focused on models, methodologies, and implementation techniques to efficiently establish the communication infrastructure (connection between communicating parts) within an embedded system.

The following Section 1.1 outlines embedded systems, their context, and motivates research of this subject. It is explanatory shown that slightly modified system specifications strongly influence the required communication infrastructure of the system implementation. Finally, Section 1.2 gives an overview of the remaining chapters of this thesis.

1.1 Embedded Systems

In contrast to general purpose systems such as personal computers, embedded systems are tailored to a certain application (see Figure 1.1) [GVNG94, Mic96, BCO96, Ern97b, Ern97a]. They keep tight coupling with their immediate (technical) environment by using sensors and actors. In general, an embedded system comprises an arbitrary number of cooperating components from different manufactures that jointly implement the specified behavior [Ern98]. The most important components include general and special purpose processors, programmable and dedicated hardware units, as well as memories. Furthermore, an appropriate implementation has to

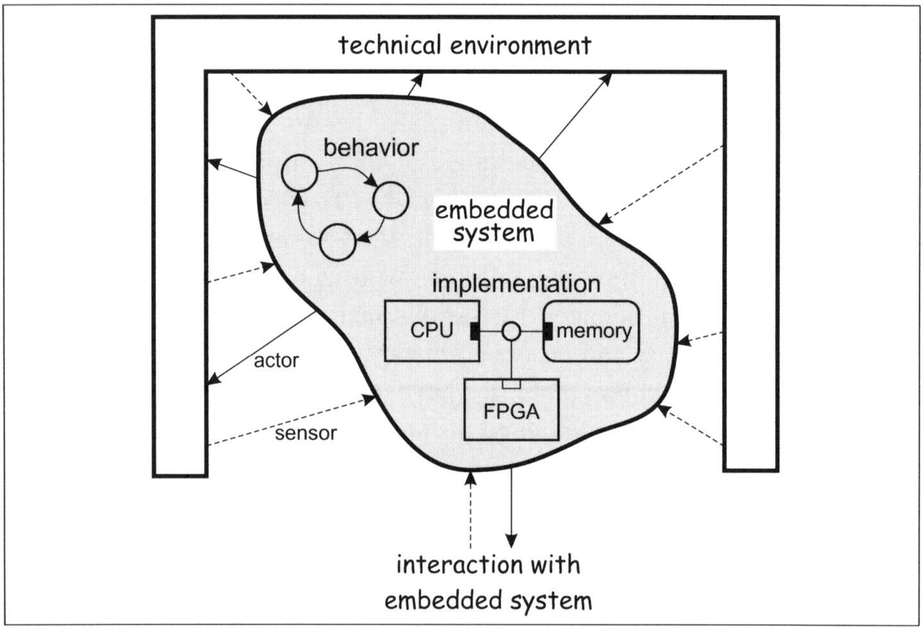

Figure 1.1: Context of an embedded system.

satisfy a number of constraints such as low power consumption [FGSS98] and real-time requirements [HBKG98]. Generally, there exists a set of alternative implementations for an aspired system behavior. They distinguish in features like, e.g., system speed and cost [BTT98] and especially by a different communication infrastructure. To control the design complexity, embedded systems are developed by the help of tools such as COSYMA [EHB+96] and PTOLEMY [BHLM91]. Nevertheless, the heterogeneity of the system components as well as different communication requirements of alternative implementations are rather challenging. For that reason, most of these tools are restricted to a certain embedded system platform.

In the context of this thesis, the primary focus of interest are issues concerning the *point-to-point communication* between interacting parts on heterogeneous targets. To motivate the research the simplified specification of the embedded system shown in Figure 1.2 is considered. Here, the behavior is described in form of a *problem graph* PG. The problem graph consists of tasks (e.g., node w) and buffers (e.g., node b_1) that communicate along directed edges. The prospective target is captured by an *architecture graph* AG. The architecture graph is a structural representation of the implementation and includes components (e.g., computing resource CPU, and memory mem) that are connected to buses (e.g., node bu_0). Components have interfaces that enable the communication among them. Finally, a *mapping* M specifies which component of the architecture graph will implement which node of the problem graph. Subsequently, an interesting

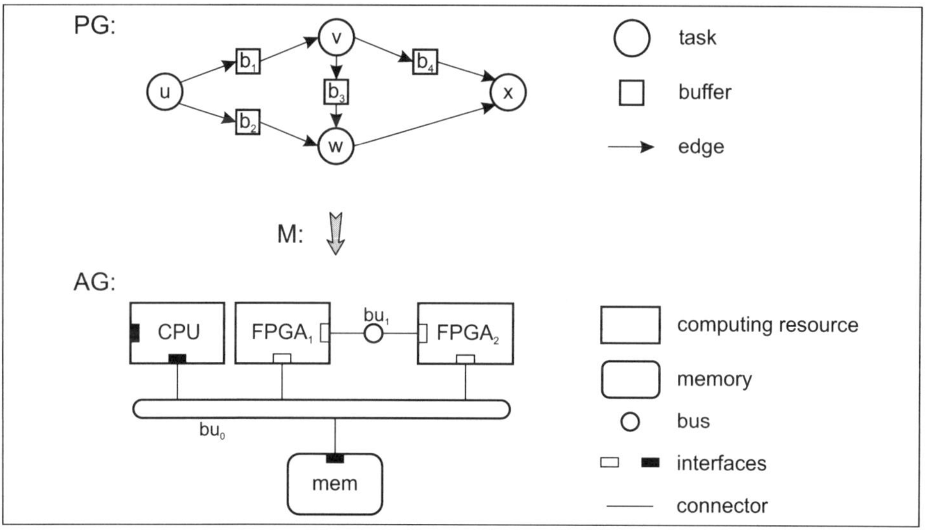

Figure 1.2: Simplified specification of an embedded system.

and challenging question during the design and synthesis process is to find an appropriate mapping M that satisfies additional system constraints such as fast execution speed and low communication overhead. To outline the complexity of this task, the following three cases are considered:

Case 1: *All problem graph nodes implemented by the same computing resource*

This is the simplest case. All problem graph nodes use the same implementation language. For example, for a processor computing resource the language C is appropriate. From a programming point of view, the communication between the nodes takes place on the same chip (i.e., *on-chip communication*). Subsequently, communication channels have to be established that are specified by the edges of the problem graph.

Case 2: *Problem graph nodes distributed among the architecture components*

Generally, the node implementation languages differ. Assume that task u has to be executed by the processor computing resource CPU and the remaining nodes are located on the FPGA computing resource $FPGA_2$. Subsequently, the communication channels between the nodes u, b_1, and b_2 make use of the bus bu_0 (i.e, *off-chip communication*). Their implementation requires appropriate *code fragments* on both computing resources to control the data transmission via the component's interfaces. In a refined problem graph, this can be captured by additional *communication nodes* as shown in Figure 1.3a). However, if it is assumed that all buffers are located

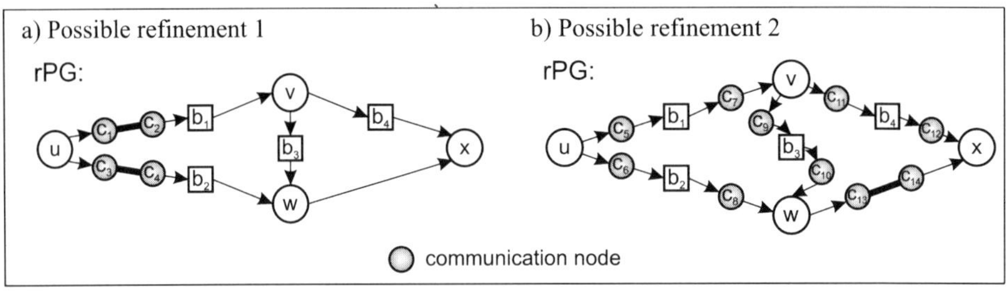

Figure 1.3: Possible refinements.

in the memory mem an appropriate refinement is shown in Figure 1.3b). Note, that depending on the location of the task nodes the implementation of the communication nodes (i.e., the code fragments) is different.

Case 3: *Dynamic reconfiguration of FPGA computing resources*

Recently, run-time reconfiguration of FPGA computing resources has arisen to optimize the system implementation [HW95, SSH+99]. This methodology involves additional control structures in the problem graph that require further communication channels. In Figure 1.4a) the considered problem graph has been extended by a control structure consisting of nodes ρ, σ_1, and σ_2 that enables run-time reconfiguration [EP02]. The nodes u and ρ are located on the CPU, the buffers are assigned to the memory mem. The configurations δ_1 and δ_2 denote a set of exclusive nodes that are implemented alternately on the computing resource $FPGA_1$. As the refined problem graph in Figure 1.4b) shows, the communication infrastructure for this kind of specification will be rather different from the last cases. Note, that communication channels can be physically interrupted during reconfiguration (i.e., *interconfiguration communication*).

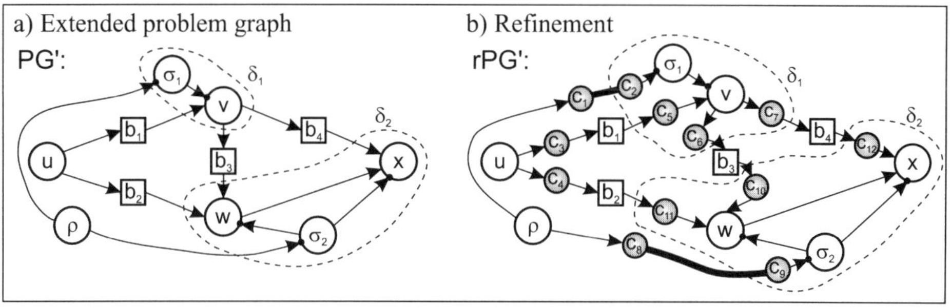

Figure 1.4: Specification of reconfigurable parts.

The above considerations outline the complexity and diversity of establishing a communication infrastructure for embedded systems. This research work provides models and methodologies to (i) capture a problem description, (ii) to apply refinements considering communication issues, and (iii) to implement an appropriate communication infrastructure.

1.2 Overview

This thesis is organized into seven chapters:

Chapter 1 ...
... outlines the problem statement and motivates the research work.

Chapter 2 ...
... considers *related work* concerning design and implementation of embedded systems.

Chapter 3 ...

... suggests *specification models* to capture and describe problem specifications that use point-to-point communication.

Chapter 4 ...

... presents appropriate *model refinements* as well as a taxonomy of arising communication types.

Chapter 5 ...

... shows *optimization and synthesis* methodologies for the suggested models. This includes a description of the design flow and a related framework.

Chapter 6 ...

... summarizes the new research results and outlines perspectives concerning further work on communication channel synthesis.

Furthermore, the **Appendix** gives a brief summary about the published *papers* and the *CV* of the author.

Chapter 2

Embedded Systems

This chapter considers related work concerning design and implementation of embedded systems. Essentially, embedded systems are dedicated to a certain environment and provide a specific behavior. Their implementation considers a number of constraints such as adapted computing performance, low system cost, low power requirements, and short time-to-market. Quite obviously, the diversity of application areas and constraints can not be covered by one universal platform. Instead, a lot of different *target architectures* have emerged that are optimized for certain application scenarios. Usually, an architecture comprises a bunch of communicating components jointly implementing the system's behavior. Each component needs interfaces and device drivers for communication. Subsequently, overhead arises that influences system performance and system cost. Therefore, *communication* as part of the behavior description as well as between target components is a major issue. Finally, the embedded system's *designflow* incorporating the above considerations is of utmost importance for an efficient implementation regarding human and technical resources.

The remaining parts of this chapter present related work concerning up-to-date *target architectures* (Section 2.1), modeling and synthesis of *communication* between system parts (Section 2.2), as well as embedded system's *designflow* (Section 2.3).

2.1 Target Architectures

Since their beginning, embedded systems consist of a set of communicating components each providing a dedicated functionality. While this concept of connecting components has not much changed the individual components have become considerably more powerful. The rapid technological progress concerning transistor density and optimized on-chip circuit architectures enabled components consisting of millions of transistors providing a remarkable processing power. Nevertheless, this dramatic development faces designers with new challenges of producing reasonable designs taking advantage of this increased silicon capacity and still reducing time-to-market. Subsequently, system integration on a single chip (system-on-chip, SoC) is aspired [LRV+96].

Regarding this thesis, a focus of research is on the interaction of components, either on-chip or off-chip. Figure 2.1 outlines a nomenclature of

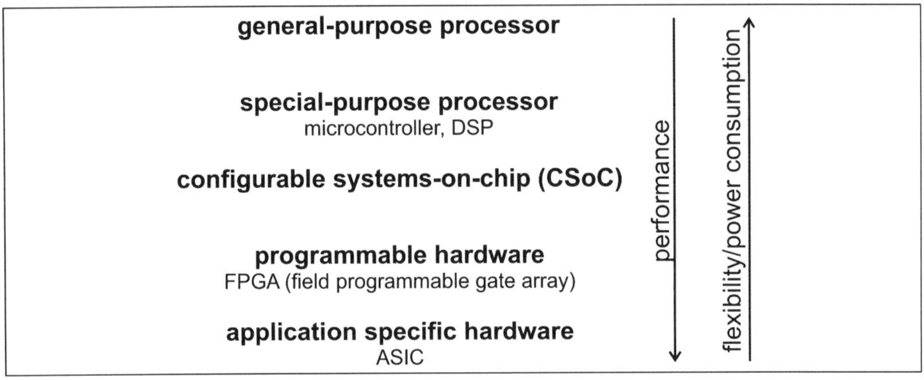

Figure 2.1: Computing components for embedded system architectures.

common computing components that are used for today's system implementation. The *general-purpose* processor provides the largest flexibility concerning its application domain but on the other hand has a high power consumption. On the other end, *application specific hardware* provides the best computing performance but is very inflexible concerning specification changes. Between these two extremes components combine different features to optimize an implementation.

General-purpose processor
A general-purpose processor provides high computing performance for a broad range of application domains [Int, Idt]. However, it is not optimized for special tasks such as digital signal processing. Therefore, it is usually used in conjunction with further computing components and implements system parts that are subject of frequent changes.

Application specific integrated circuit (ASIC)
ASICs are application specific components used to implement and speed-up performance critical parts of an embedded system [SYN]. They have the best $performance/power$ ratio and quite often represent a complete SoC. However, their development is time and cost intensive. Nevertheless, they often emerge as the cheapest alternative for high volume products.

Special-purpose processor
SoC approaches tend towards large initial cost. Therefore, processor vendors such as Motorola [Mot], Philips [Phi], and Texas Instruments [Ins], provide microcontroller and DSP (digital signal processor) families. Microcontrollers are optimized for control-flow applications and consist of a processor core surrounded by a set of peripheral units such as dedicated I/O facilities and on-chip memory. DSPs are optimized for dataflow problems, can have fast on-chip data/program memory and provide a high data throughput.

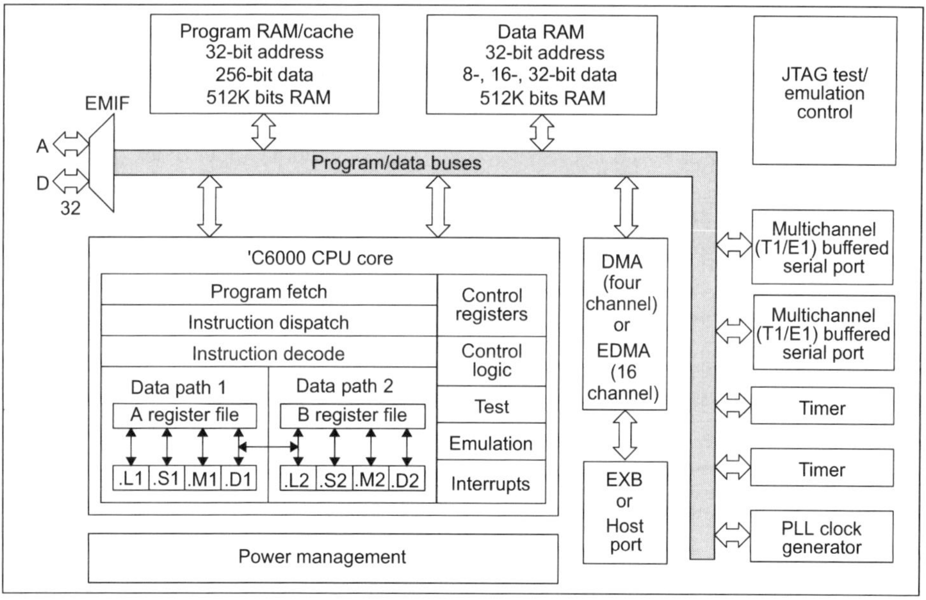

Figure 2.2: TMS320C62x/C67x Block diagram.

Figure 2.2 outlines the block diagram of an up-to-date high performance DSP family, the Texas Instruments TMS320C62x/C67x. It consists of a CPU core, on-chip memory, and peripherals. Family members provide fixed or floating point arithmetic and differ in their memory configuration and on-chip peripheral modules.

Programmable hardware
In 1985, XILINX introduced the first family of FPGAs (Field Programmable Gate Arrays) representing a trade-off between the large computing power of a dedicated hardware implementation, reprogrammability, and cost. Currently, FPGAs are available (XILINX Virtex-E series) comprising million of system gates, supporting 20 different input/output standards at a speed of up to 300 Mbits/pin, and as well include dedicated on-chip RAM. Such components enable to build complex SoC systems such as Adobe PhotoShop filters [LSS99], DES encryption [Pat00], and network processors [LNTT01].

Figure 2.3 shows the architecture of the state-of-the-art XILINX Virtex-E series consisting of an array of configurable logic blocks (CLB), routing resources (e.g., VersaRing), large blocks of on-chip RAM (BRam), and I/O blocks (IOB). CLBs implement combinatorial/sequential logic and provide additional carry logic for fast arithmetic function implementations. Input/output blocks connect the programmable area with chip pins.

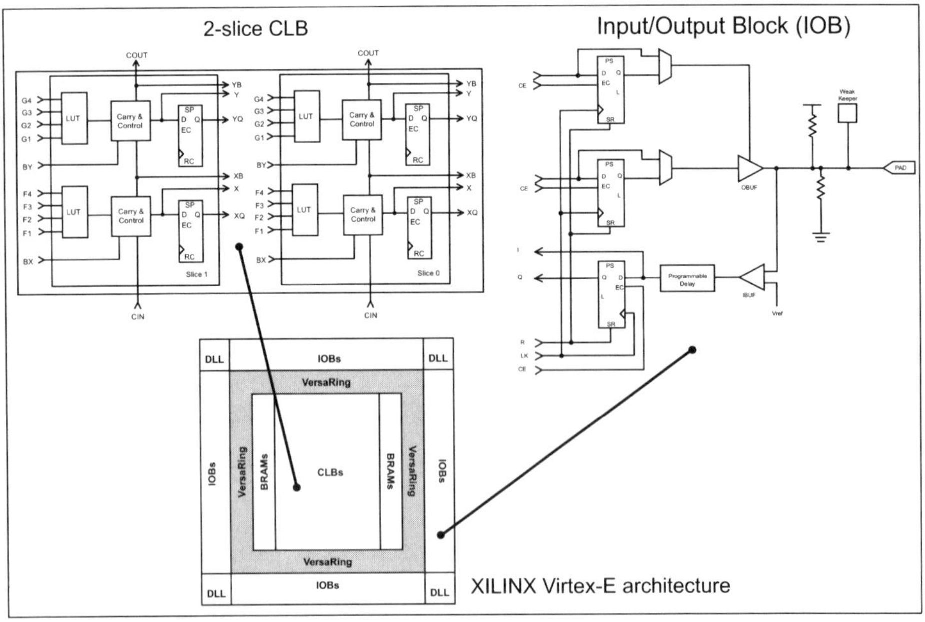

Figure 2.3: Architecture of the XILINX Virtex-E series.

Configurable systems-on-chip (CSoC)

Recently, configurable systems-on-chips have emerged. These components exploit the advantages of software flexibility and the speed of a dedicated hardware implementation. Usually, they consist of a general-

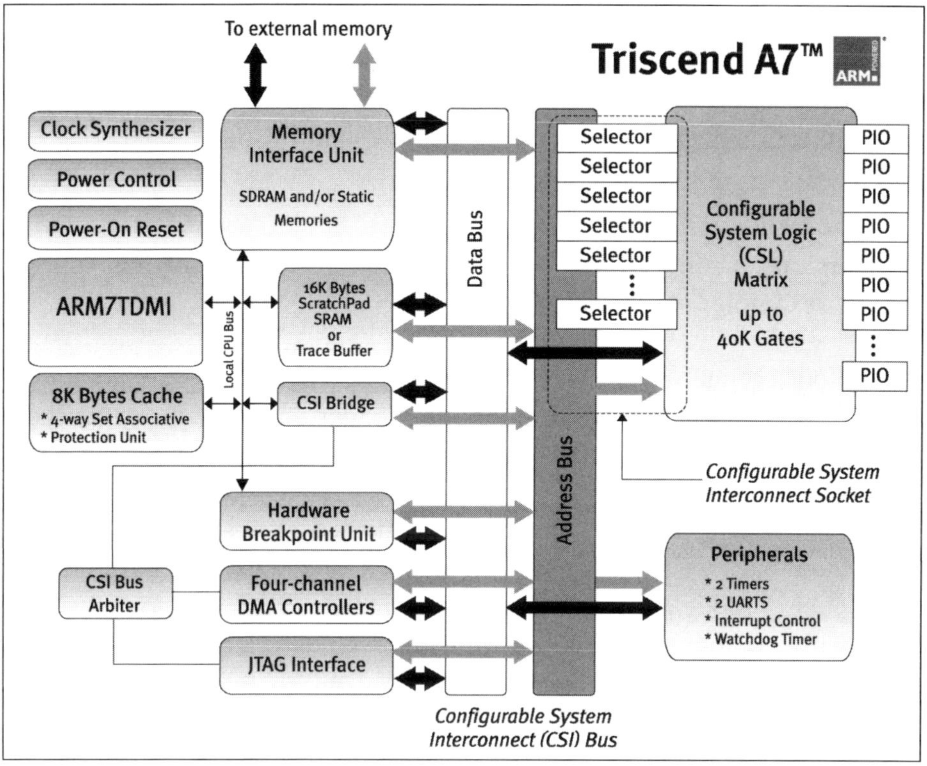

Figure 2.4: Triscend A7: Configurable SoC (by courtesy of Triscend Inc.).

purpose processor core, a configurable logic core, and memory [Kea00]. Related research projects include the NAPA project [RLG$^+$98], or DReAM [BG00] a SoC platform intended for mobile communication. Commercial solutions include the Excalibur processor solution from Altera [Exc] integrating RISC processor cores (ARM, MIPS (hard cores) or Nios (soft core)) into their programmable components, the Triscend A7 [Tri] CSoC combining an ARM7 core with configurable logic (see Fig 2.4), or the co-operation between XILINX and IBM [XILb] integrating a PowerPC core with configurable logic.

2.2 Communication

Data communication is a performance critical issue for embedded systems. It affects communication speed, system cost, and power consumption. Not only the intricate relationship with the environment is important (i.e., by measuring/observing physical parameters using sensors and appropriate reaction using actors) but also the local communication between processors, custom hardware, and memories has a large influence on the overall computational power. The complexity of embedded systems concerning communication between components has two main issues:

Communication semantics
Different specification formalisms have different communication semantics. For example:

- Nodes of an SDF [LM87b] graph fire instantaneously if enough tokens are available on all node input edges and wait for data otherwise. Edges have an indefinite buffer capacity.

- In FunState [Str00] a function is activated by a state machine and then reads/writes its inputs/outputs instantaneously. A buffer is explicitly modeled. Read access to a register buffer is non-blocking which may potentially lead to invalid values whereas a read from a queue buffer is blocking.

Implementation techniques and devices
The implementation of the infrastructure providing data communication is influenced by factors such as performance requirements, architecture component types, supported communication protocols and synchronization techniques, and interconnection scheme. For example:

- Operating systems like, e.g., Virtuoso [Eon99] for DSPs, provide a set of different communication mechanisms. By using asynchronous events and semaphores, tasks can notify each other if there is a block of data within a shared memory region to be transferred. More comfortable, mailboxes and FIFOs allow an easy message transfer providing asynchronous or synchronous communication. As a consequence, they cause a degradation in communication performance.

- Future aspects like, compatibility, may influence the implementation. For example, the decision to use an off-the-shelf PCI controller that is compliant with the whole PCI local bus specification or to implement only the required subset of the PCI specification may lead to compatibility problems during product maturation. Where in the first case compatibility of the embedded system is ensured with all PCI systems and the overall design time is reduced the second solution may be much more cost efficient for large volumes and less power consuming if only a small subset of the PCI specification is required.

Due to the outlined reasons above, specification models that target system implementation include a model for communication [COB92, LV94, DMBIJ97, BTT98, HBKG98, PRSV98, EP00c].

2.2.1 Modeling Communication

Including communication in a specification model has several consequences. Besides the pure notation of elements of the communication model the following issues have to be considered:

- Communication properties provided by a target depends on the particular implementation (architecture components, supported protocols, bus structure, etc.). Subsequently, an accurate estimation concerning communication overhead is very difficult in early design stages [YW95, KM98].

- Major constraints on binding and allocation are implied as not all combinations of communication methodologies are supported between the connected components.

- Storing interface implementations for all possible combinations of computation components is not feasible [ETT98].

- Programmable hardware components like, FPGAs, originally do not provide any kind of interface implementation [ET98a].

A specification that considers the above issues includes three parts: (i) the specification of *necessary* communication channels between interacting units, (ii) the specification of *available* communication resources, and

(iii) a mapping between necessary communication channels and available communication resources [JO95, VRBM96, RSV97, OB98, EP00c]. Usually, these parts are included in a *behavior model*, an *architecture model*, and a *mapping* between behavior and architecture model respectively.

Behavior model
Modeling communication in a behavior model denotes the necessary data channels between communicating objects. It is represented, e.g., by communication nodes and directed edges in a problem graph [TBT97, YW95], using dedicated language constructs [Gon97] within the functional description, by communication primitives providing dedicated semantics [JO95, DIJ95, VRBM96, LRV$^+$96, VT97, RSV97, EP00c], etc.

Architecture model
The *available* communication resources are specified in an architecture model. Resources include bus connections between components [TBT97, EP00c], routing capabilities of computing resources [OB98, EP00c], available I/O modules of computing resources and memories [EP00c], supported communication protocols, etc.

Relating necessary communication paths with available resources
For the later synthesis process the available communication resources have to be assigned to necessary communication channels. For example, in [TBT97] communication nodes are assigned to bus resources of the architecture model. In [OB98, EP00c] a communication path is assigned to each directed edge in a problem graph.

Specifications comprising these three parts often provide estimations about the implementation of the communication such as the expected area overhead [KM98, FSS99, EZT00]. In this monologue, the investigations are restricted to the implementation of point-to-point channels although there potentially exists broadcasting (i.e., one sender sends a message using one logical channel to several receivers). However, broadcasting can be easily modeled. A sender emits its message to a "distributor" that duplicates the message and forwards it to a set of connected receivers.

2.2.2 Communication and Interface Synthesis

The goal of communication and interface synthesis is to establish an appropriate *communication infrastructure* based on the available resources of the architecture. This infrastructure provides the necessary communication channels between interacting objects. Therefore, the communication channels modeled in the behavior model are stepwise transformed into device drivers for processors and interface circuitry for hardware programmable resources [RSV97, LRV^{+}96, OB98, EP00c]. Various methodologies have emerged to establish a communication infrastructure. Quite often, several methodologies are used concurrently.

By hand
Writing device drivers and interface circuitry by hand is still the most applied technique because of the many standards and proprietary connection schemes. Besides that, the remaining approaches quite often require some degree of manual work as well.

Library based
Library based approaches provide predefined and pre-tested device drivers and interface circuits stored in a library that are instantiated and configured during system's compile time. This technique is usually employed if the target platform remains fixed. Examples include the investigation of HW/SW partitioning algorithms for a certain target platform [DMBIJ97, CV99] and commercial user programmable embedded systems where the manufacturer delivers a corresponding API (e.g., Sundance [Sun]). Library based co-design frameworks include, e.g., COSYMA [EHB^{+}96], Akka [TOJH96], Vahid/Tauro [VT97], DICE [HBKG98], CCS [ET98b].

Template based
Templates are predefined interface codes (e.g., written in C and VHDL) stored in a library where (i) tags have to be replaced during compile time by real code fragments, or (ii) code fragments are stored in a library that are used to compose a real driver/circuit at compile time. This technique provides greater flexibility than pure library based approaches. However, it has potentially more sources for error as codes of various libraries may be combined. For example, using *#define*'s in C-header files provides a simple technique to adapt a device driver to a target, e.g., to set the base

address register for I/O module access. Template based co-design frameworks include, e.g., CCS [ET98a].

Pattern based
Patterns [GHJV95] describe a common problem that occurs over and over again and describe the core of the solution to that problem in such a way that the description may be reused many times [Ris98]. Patterns for communication and interface synthesis describe typical connection problems for architecture components but obviously requires manual work to establish the actual communication infrastructure. Nevertheless, it is an efficient way to reuse existing knowledge to create new device drivers and interface circuits. An approach in this sense are general protocol descriptions, e.g., that describe how to access a shared memory region, without concrete suggestions concerning the implementation.

Component based
A component (often called IP (**I**ntellectual **P**roperty) core) is a predefined implementation of a computational unit, e.g. a JavaBean [Mic] (software) or an FIR filter (hardware). It provides a standardized interface that allows simple intercomponent connection. Component based design is gaining interest as it enables to rapidly create dedicated SoCs of large complexity [ZG97, COH+99, EPT99, GDZ99]. For this reason, companies like such as Mentor Graphics [Gra] and SYNOPSYS [SYN] provide IP libraries from simple registers to complete processors to be used in hardware designs. However, this approach rises problems concerning the interconnection between these IP components. Only few (quasi-)standards for hardware components have emerged yet. Examples include the *VCI* (Virtual Component Interface) standard of VSIA (Virtual Socket Interface Alliance) [VSI] and the *OCP* (Open Core Protocol) of Sonics Inc. [Web00]. Therefore, connecting IP cores still makes a considerable amount of manual work necessary. Few commercial tools exist providing the integration of IP cores like, Cierto VCC (Cadence) [Cie] or COSSAP (Synopsys) [COS]. Related research approaches include [VLM96b, VG98, COH+99]. Recently, programmable IP cores [RST+00, Röw00] have been suggested providing a programmable embedded processor to enhance the flexibility of the cores in terms of supported interfaces, new core function releases, and built-in self test. However, this flexibility is paid with additional area overhead.

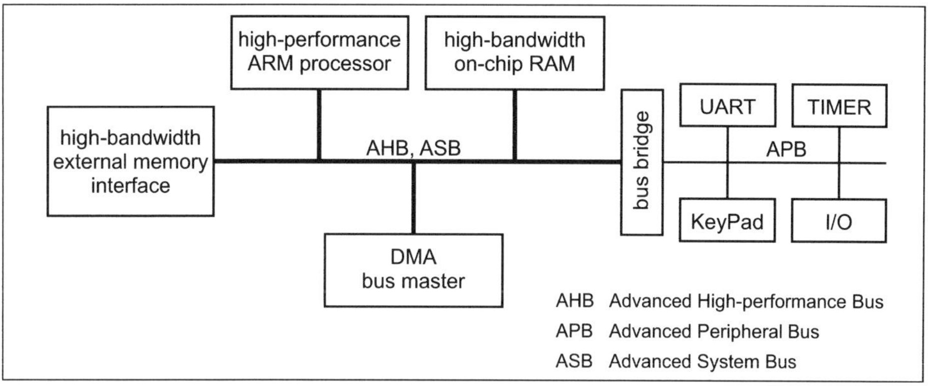

Figure 2.5: AMBA, Advanced microcontroller bus architecture [ARM].

Generator based

Due to the diversity of architecture components and IP cores only few approaches arose that automatically establish the complete communication infrastructure. Known approaches base on the formalization of component timing diagrams [Bor92, COB92, COB95a] or signal transition graphs [LV94] followed by the synthesis of appropriate finite state machines and glue logic [Bor88]. With the appearance of IP cores for SoC design, algorithms for transducer synthesis are gaining interest that allow to establish point-to-point channels between communicating IP cores [RSV97, OLB98, PRSV98]. Commercial approaches like, e.g., Cierto VCC [Cie], often provide a mix between library based and generator based approaches to synthesize intermediate transducers. Predefined communication primitives (FSMs) are stored in a library that can be modified, simulated, and finally synthesized. The framework CCS [ET98a] supports generator based approaches by the use of appropriate *plug-in generators* that encapsulate such techniques.

Platform based

To reduce the SoC design complexity pre-designed SoC architectures have arisen that are adaptable for specific application requirements [FSV99, Rab00, ARM, GV00]. They are equipped with programmable general and/or special purpose processor cores, communication buses, memory controllers, on-chip memory, I/O controllers, etc., and provide a time efficient way to implement an application by reducing system cost and development time. Platform based design may include the use of wide spread bus standards such as the VME bus for backplanes [IEE87], the PCI bus for PC based systems [PCI95], or the CAN (Controller Area Network) bus for automobile industry [ELSS94]. As an example, AMBA (Advanced Microcontroller Bus Architecture) is a high performance SoC communication standard developed by ARM [ARM] (see Figure 2.5). Heart of such a system are two communication buses connected via an intermediate bus bridge. A high-performance/system bus provides the connection of processor cores, and high speed RAM. Peripherals are connected to the much slower peripheral bus. The standard stipulates bus protocols and component macros that enable an easy connection of components.

2.3 Designflow

The *ideal* designflow to develop an embedded system starts with a pure behavior description of the problem and considers constraints such as cost, real-time behavior, and power consumption. Conflicting design objectives such as cost versus computing power [BTT98], hardware/software partitioning respecting communication [CV99], or FPGA run-time reconfiguration [HW95] respecting computation power are considered using design space exploration on different levels of abstraction. The specified system behavior is refined (either full or semi-automatically) towards an implementation considering alternative solutions and finally results in an implementation consisting of connected electronic components jointly executing and implementing the specified behavior. The importance and usefulness of full- or semi-automatic tools for mapping high-level specifications onto heterogeneous target architectures has been recognized by many research groups [PL91, KKR94, BR95, HBK96, EHB$^+$96]. However, to develop a framework in the sense of the ideal designflow the following two important issues arise:

Dedicated specification formalisms for different application domains
The specification formalism covering all known application domains has not been found yet. Instead, formalisms dedicated to specific problem domains have been developed. For example, controlflow oriented systems can be described by, e.g., Harel Statecharts [Har87] and finite state machines. Dataflow oriented systems are described and implemented by using models such as SDF (synchronous dataflow) [LM87a, LM87b, Zep95],

BDF (boolean dataflow) [Buc93], and cyclostatic dataflow [BWE+93], [EBLP94]. Recently, frameworks such as MOSES [Jan00] and Ptolemy [BHLM91] have arisen, dealing with modeling and simulation of embedded systems using different formalisms. They provide a generalized inner computation model and even allow an implementation on heterogeneous target architectures [EZT99].

Component heterogeneity of the target architecture
The diversity of available architecture components and their rich set of interfacing methodologies makes it difficult to create a fully automatic tool. Especially tools, that support communication and interface synthesis for various target components are rare. However, such tools are a prerequisite for efficient design space exploration and rapid prototyping. A main reason for this lack of tool support is the abstraction level of the architecture components. For example, a sophisticated processor model as used in [BEK+95, BCG+97] allows an efficient implementation for a single processor or a dedicated hardware technology. But, retargeting and taking account of other processors or programmable hardware components is not possible. On the other hand, a simple model providing just a node in an architecture graph without fine grain hardware information allows a sophisticated algorithmic study of complex design problems but prevents an efficient implementation due to the high abstraction level. Therefore, an appropriate target component model useful for system design and implementation is a trade-off between model granularity and retargeting flexibility [ETT98].

Unfortunately, most frameworks for embedded systems lack a fast and flexible back-end for the final code generation that incorporates communication and interface synthesis for heterogeneous platforms. Current frameworks are focused on domain-specific approaches such as the Polis environment for control-dominated systems aiming at a processor-coprocessor architectures [BCG+97], or the CoWare approach [VLM96a, MBL+96, VRBM96] for digital signal processing problems targeting at system-on-a-chip solutions . Further related research approaches include:

VULCAN-II [GM93] ...
... a hardware oriented approach that starts with a behavior description in HardwareC (C like syntax but hardware semantics). To reduce the implementation cost non-time-critical parts are moved from a hardware im-

plementation to a software implementation on a standard processor or a processor core. A shared memory connected to a global system bus is used to implement the communication between processor and custom hardware.

COSYMA [EHB$^+$96] ...
... a software oriented approach for small embedded real-time systems. A behavior specification is written in C^x a superset of C. A simulated annealing algorithm starts with the complete application in software and interactively moves software parts to a co-processor until all constraints can be met. Its target consists of a Sparc processor and an FPGA board used as co-processor. Communication between processor and co-processor is implemented using a shared memory.

Ptolemy [BHLM91, Pto] ...
... an interactive framework for simulation, prototyping and software synthesis of DSP systems. It supports several models of computation, e.g., SDF, dynamic dataflow, discrete event, etc. The behavior is manually partitioned into software executed on DSPs and hardware implemented as custom datapaths or on FPGAs.

DICE [HBKG98] ...
... starts with an arbitrary number of concurrent C and VHDL processes that are converted into a CDFG (control dataflow graph). The interactive partitioning approach [HBKG98] tries to keep as much as possible in software. Hardware processes are generated automatically if nodes of the CDFG are moved to a hardware implementation. The target consists of a set of connected modules, i.e., microprocessors, DSPs, ASICs, and FPGAs.

SIERA [SB95] ...
... maps a network of concurrent processes onto a printed circuit board specified as an architecture template comprising FPGAs, ASICs and processors. Reusable hardware and software modules are stored in a library and are connected via a layered interconnect model and predefined hardware/software primitives for low-level message passing between pairs of processors.

Chinook [COB95a, COB95b] ...
... maps a behavioral description of communicating processes onto a single or multiprocessor system and generates software drivers and glue logic to connect processors and external chips. FPGAs are not supported.

The approaches discussed above implement a certain designflow to transform the initial behavior specification onto a target. Besides the different aimed targets and problem domains, hardware/software partitioning has a central role as it strongly influences the implementation. The following two sections outline the basic designflows used in research and industry favoring *early* or *late partitioning* respectively.

2.3.1 Conventional Designflow

A conventional designflow is a rather rigid sequence of transformative steps [BR95]. It starts with the analysis of requirements and constraints (design capture) leading to a corresponding integrated system implementation (see Figure 2.6). Hardware/software partitioning is an early design step that divides a problem specification into sub-specifications for software and hardware. Subsequently, both parts are developed separately. Interface synthesis establishes the communication infrastructure consisting of device drivers and interface circuitry by using the connected components of the target architecture. During a final integration step the individual elements are assembled, prototyped, and tested. Unfortunately, this design style is strongly shaped by the *early partitioning*. This often prevents rectifying redesign loops and design space exploration due to expensiveness and rigid time-to-market constraints. Nevertheless, it is broadly applied in industry as its straight development process allows an easy management of human, technological, and time resources. Related research approaches that support a conventional approach include, e.g., VULCAN-II [GM93], COSYMA [EHB$^+$96], DICE [HBKG98], SIERA [SB95], and Chinook [OB98]. However, all of them include redesign loops to find optimal solutions.

2.3.2 Model-based Designflow

In contrary to a conventional approach a model-based designflow does not propose a fixed sequence of refinement steps [BR95]. It rather bases on a

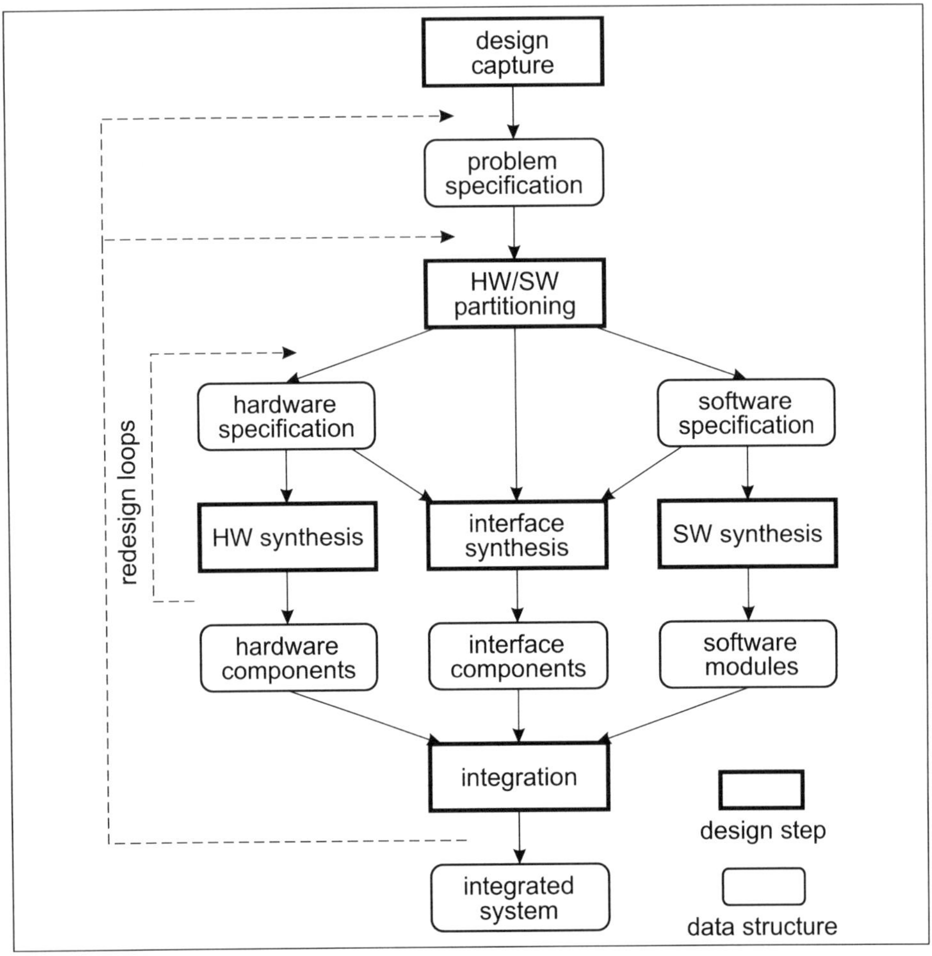

Figure 2.6: Conventional designflow.

set of tools that work on a shared repository of dynamic data objects (see Figure 2.7). Tools can be executed in different orders to evaluate and explore alternative target architectures considering important aspects such as different computing resources, protocol selection, communication overhead, implementation speed, and estimated target price. As the tools are not a fixed part of an overall designflow new tools can easily be integrated into such a framework, e.g., to support reconfigurable embedded systems [EP00a]. Essentially, no distinction is made between hardware and software function implementations in early design phases dropping the weak point of conventional design approaches. Instead, the approach favors a

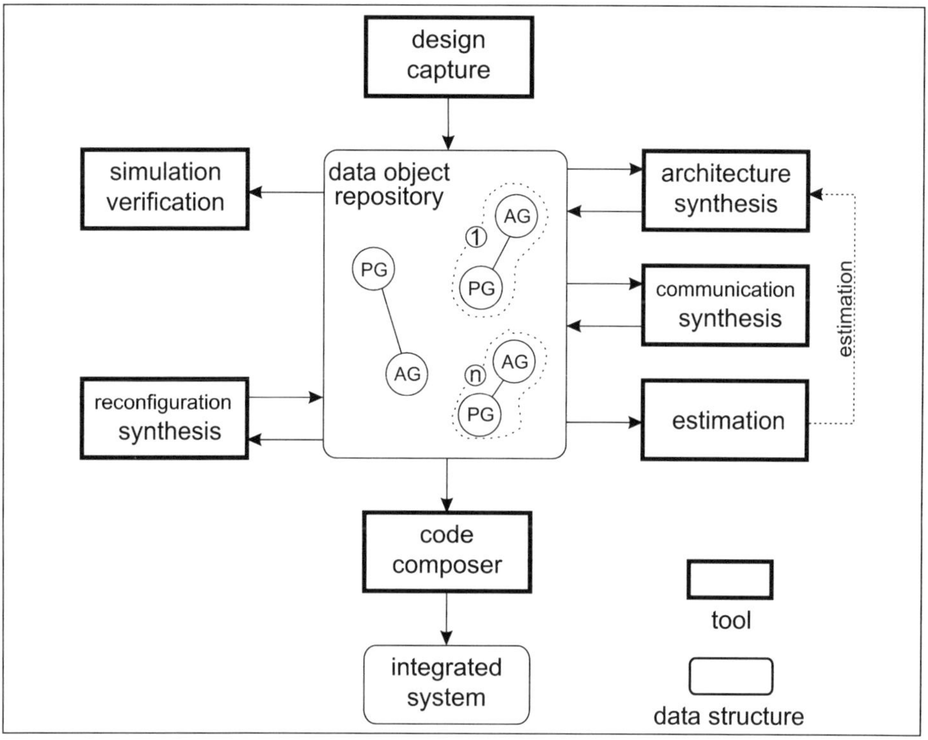

Figure 2.7: Model-based co-design methodology.

late partitioning and is able to find better implementation solutions. Different implementations are evaluated based on early estimates trading-off program/data memory, execution time, etc., leading to a concurrent design style using the synergies of hardware and software implementations [EZT99].

As visualized in Figure 2.7, the heart of such a framework is a *data object repository* that captures the problem specification and prospective solution objects on various abstraction levels. *Design capture* tools allow to formulate a design problem as a data object comprising a behavioral description of the system's functionality, its requirements and constraints, e.g., in form of a *problem graph* (*PG*). *Architecture synthesis* tools seek for optimal implementations and allocate target components to define the target architecture in form of an *architecture graph* (*AG*). *Simulation and verification* tools allow to simulate, validate, and verify a selected implementation. *Communication synthesis* tools establish the infrastructure for communication links between connected target components. They automatically generate the required interface circuitry and software drivers based on information about the connected architecture components [EPT99]. *Reconfiguration synthesis* tools enable run-time reconfiguration of programmable hardware resources [EP00a]. *Estimation* tools allow to assess a data object, e.g., the estimated overhead in terms of hardware area required for the communication infrastructure. Finally, *code composer* tools assemble the codes to establish the integrated system. Related approaches include MOSES [Jan00], CCS [EZT99], and Ptolemy [BHLM91].

Chapter 3

Specification Models

This chapter, as well as the following two chapters describe the proposed design methodology for embedded systems (see Figure 3.1). Chapter 3 introduces specification models to capture a problem, Chapter 4 describes appropriate model refinements, and Chapter 5 elaborates the general synthesis process.

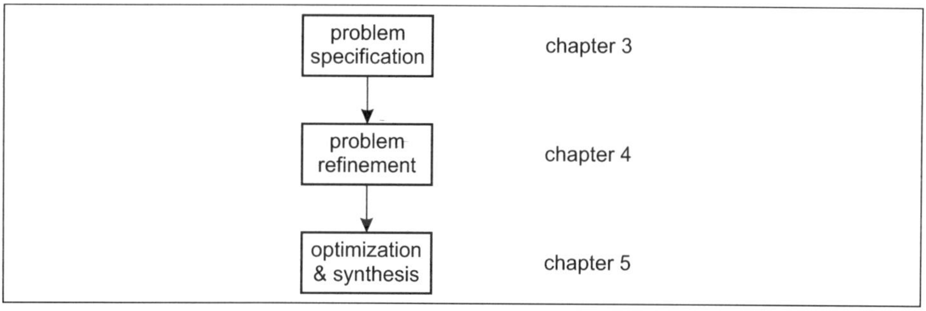

Figure 3.1: Simplified design flow.

Essentially, specification models provide formalisms to capture and describe a synthesis problem. In that sense, the proposed methodology comprises a hierarchy of formalisms of various modeling capabilities targeting at dataflow oriented heterogeneous systems (see Figure 3.2).

The *GPS formalism* (**G**eneral **P**roblem **S**pecification) ...

... is parent of all proposed specification formalisms and provides definition of terms, basic elements, and composition rules to specify a dataflow problem for synthesis. The principal focus of interest is on the specifica-

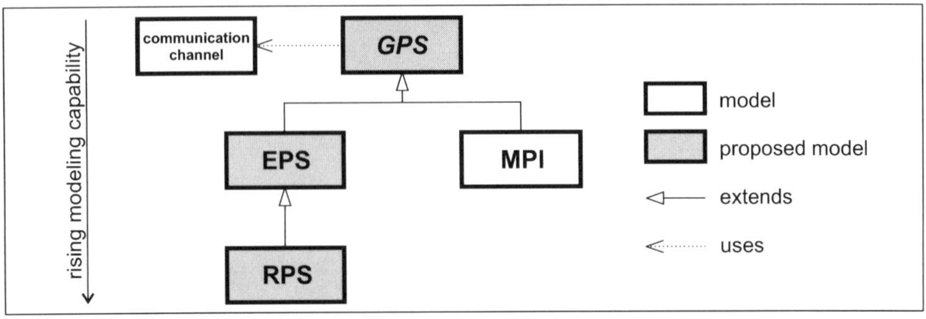

Figure 3.2: Model hierarchy (UML notation [PJ99]).

tion of communication channels between communicating system parts that base on the model of a *point-to-point communication channel* (see Section 4.1).

The *EPS formalism* (**E**mbedded system **P**roblem **S**pecification) ...

... is an extension of GPS. It aims at the specification of dataflow problems for embedded systems composed of software and hardware programmable devices as well as memories.

The *RPS formalism* (**P**roblem **S**pecification for **R**econfigurable systems) ...

... is an extension of EPS and additionally supports the specification of systems containing dynamic reconfigurable hardware devices [EP00a].

The remaining parts of this chapter introduce the three proposed specification formalisms GPS (Section 3.1), EPS (Section 3.2), and RPS (Section 3.3) accompanied with explanatory examples. A short summary in Section 3.4 concludes this chapter.

3.1 General Problem Specification (*GPS*)

The GPS formalism is the foundation for the specification of synthesis problems for dataflow oriented heterogeneous systems. It provides definition of terms, basic elements, and composition rules. Hence, GPS can be viewed as parent of models that aim at the specification of synthesis problems for models such as, marked graphs [CH71], Kahn process networks [Kah74], synchronous dataflow graphs [LM87a, LM87b], FunState components [TSZ$^+$99], and MPI [Pac97]. However, GPS is not restricted to embedded system specifications only. As an example, the definition of an *MPI formalism* (**M**essage **P**assing **I**nterface) [Pac97] would enable to specify synthesis problems for networks of distributed workstations. In terms of UML (**U**nified **M**odeling **L**anguage) [PJ99], GPS is an abstract model (cursive face in Figure 3.2).

GPS is quite similar to other approaches [SB95, Wol97, RSV97] in the sense that it separates system behavior from target architecture. It comprises three specification parts that are discussed in this chapter:

- a platform independent behavior specification in form of a *problem graph*,
- the structure and components of the prospective target platform described by an *architecture graph*, and
- a *mapping* denoting which component of the architecture graph will implement or execute which node and communication relation of the problem graph.

3.1.1 Problem Graph

The behavior of a system is captured in form of a problem graph. Essentially, such a graph is platform independent and consists of a set of communicating nodes.

Definition 1 (Node v) *A node $v = (I_v, O_v)$ has a set of input ports I_v and a set of output ports O_v.*

A node reads data from its input ports, processes these data, and writes results onto its output ports. In literature, nodes are also called *actors* [LM87a].

Definition 2 (Problem Graph PG) *A problem graph $PG = (V, E_{PG}, I, O)$ consists of a set of nodes V, a set of node input ports $I = \bigcup_{\forall v \in V} I_v$, a set of node output ports $O = \bigcup_{\forall v \in V} O_v$, and a set of directed edges $E_{PG} \subseteq (O \times I)$.*

A *directed edge* $e = (o, i)$ is an ordered pair of ports and represents flow of data between nodes. As an example, Figure 3.3 shows a simple problem graph consisting of three nodes a, b, and c and two edges e_1 and

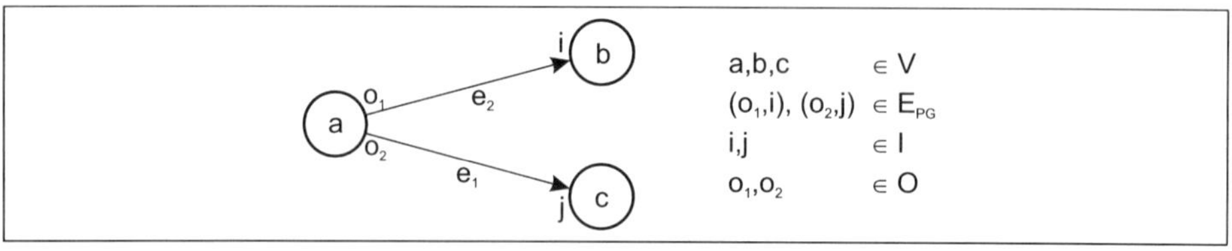

Figure 3.3: Specification of a simple problem graph.

e_2. The function $source : E_{PG} \to O$ denotes the source port of an edge. The function $sink : E_{PG} \to I$ denotes the sink port of an edge. The function $node : (O \cup I) \to V$ denotes the associated node of a port.

The communication of data via a port is specified by a set of rules. In our terminology, such a set of rules is called *protocol.*

Definition 3 (Protocol Ω) *Ω denotes the set of all protocols known to the environment.*

The function $portProtocol : (O \cup I) \to \Omega$ associates a protocol with each node port. As an example, a valid function for the simple problem graph in Figure 3.3 is:

$portProtocol(o_1) = portProtocol(o_2) = portProtocol(i) = p1$,
and $portProtocol(j) = p2$, where $\{p1, p2\} \subseteq \Omega$.

3.1.2 Architecture Graph

The prospective system target is captured by an architecture graph. Basically, it is a structural representation consisting of components, buses, and connectors. Components model units of the prospective target (like a processor in the EPS model) that implement nodes and possibly edges of the problem graph.

Definition 4 (Component c) *A component c has a non-empty set of interfaces Γ_c.*

An interface represents an input/output module of a unit of the target. Target units need *device drivers* and *interface circuits* to communicate via their input/output modules.

Definition 5 (Architecture graph AG) *An architecture graph* $AG = (C \cup B, \Gamma, E_{AG})$ *is a non-directed connected bipartite graph consisting of a set of components* C*, a set of bus nodes* B*, a set of interfaces* $\Gamma = \bigcup_{\forall c \in C} \Gamma_c$*, and a set of undirected edges* $E_{AG} \subseteq (\Gamma \times B)$*. There exists at most one edge between an interface and any bus.*

A *undirected edge* $e = \{\gamma, b\}$ is an unordered pair $\gamma \in \Gamma, b \in B$ that connects component interfaces and buses. The function $intf : E_{AG} \rightarrow \Gamma$ denotes the interface of an edge. The function $bus : E_{AG} \rightarrow B$ denotes the bus node of an edge. Usually, the set of interfaces connected to the same bus is limited. For example, a connection based on the RS232 standard has exactly two communication partners. The function $partners : \Gamma \rightarrow \mathbb{N}$ denotes the maximal number of interfaces that may be connected via the same bus. Figure 3.4 shows the specification of a simple architecture graph comprising three components cr_1, cr_2, and cr_3, two bus nodes b_1 and b_2, and edges $e_1 = \{i, b_1\}$, $e_2 = \{k, b_1\}$, ... connecting them.

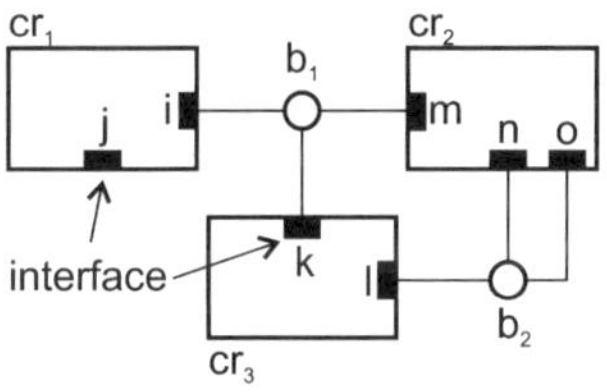

$$\{cr_1, cr_2, cr_3\} = C$$
$$\{i, j\} = \Gamma_{c_1}$$
$$\{m, n, o\} = \Gamma_{c_2}$$
$$\{k, l\} = \Gamma_{c_3}$$
$$\{b_1, b_2\} = B$$
$$\{\{i, b_1\}, \{k, b_1\}, \{m, b_1\}, \{l, b_2\}, \{n, b_2\}, \{o, b_2\}\} = E_{AG}$$

Figure 3.4: Specification of a simple architecture graph.

There exist two different types of component interfaces: *Programmable interfaces* (P-interfaces) Γ^P and *reconfigurable interfaces* (R-interfaces) Γ^R, $\Gamma = \Gamma^P \cup \Gamma^R$. The function $kind : \Gamma \rightarrow \{P, R\}$ associates a type with each interface. The function $component : \Gamma \rightarrow C$ denotes the associated component of an interface. P-interfaces model input/output modules with predefined implementations within the target unit that need a device driver for communication. Typical components with P-interfaces are general/special purpose processors (e.g., MC68340 microcontroller [Mot94]), memories, PCs (personal computers) with an expansion bus (e.g., PCI bus), etc. R-interfaces model input/output modules without predefined implementations that need additional interface circuitry for communication. This interface type provides maximal flexibility concerning the connectivity to other components. A configured R-interface behaves like a

P-interface. Components with R-interfaces include FPGAs [XILa, ALT], hardware reconfigurable systems-on-chip [Tri, Exc], etc.

The required device drivers and interface circuits for the component interfaces are the main focus of the later synthesis process as they represent the connecting link between nodes of the problem graph and target units. For that purpose, each interface provides a set of interface generators able to generate the necessary interfacing codes.

Definition 6 (Interface generators G**)** *G denotes the set of interface generators known to the environment.*

The function $generators : \Gamma \rightarrow \wp(G)$ associates a set of interface generators with an interface. It is an non-empty set in case of P-type interfaces as at least one generator is necessary to generate a device driver. The two functions *protocol1* and *protocol2* capture the connectivity of the device drivers and interface circuits produced by the generators. The function $protocol1 : G \rightarrow \Omega$ associates a first protocol with a generator. The function $protocol2 : G \rightarrow \Omega$ associates a second protocol with a generator. The properties of P- and R-interfaces are summarized in Tab. 3.1.

	P-interface	**R-interface**
implementation	predefined	requires circuit synthesis
reconfigurability	-	yes
protocols	predefined	potentially any
programming	device driver	device driver (often not required)

Table 3.1: Properties of P- and R-interfaces.

A *feasible connection* between component interfaces and buses has to satisfy the following constraints:

C1 It is assumed that n P-type interfaces $\gamma_1 ... \gamma_n$ and m R-type interfaces $\gamma_{n+1} ... \gamma_{n+m}$ are connected to a single bus b. Subsequently, it must hold that:

C1.1 $\forall i = 1...n : partners(\gamma_i) \geq n + m$.

C1.2 $generators(\gamma_1) \cap ... \cap generators(\gamma_n) \neq \varnothing$.

Comment: Constraint C1.1 *assures that all P-type interfaces connected to a single bus b support the number of connected interfaces. Constraint* C1.2 *assures that all P-type interfaces have at least one common generator which enables the communication between the connected interfaces.*

Between any two components there exists a non-empty set of potential communication paths.

Definition 7 (Communication Path Π) Π *denotes the set of all possible paths in the architecture graph. A single path $\pi \in \Pi$ denotes a path between components of the architecture graph where the elements of π are edges of E_{AG}.*

For a single path $\pi = (e_1, e_2, ..., e_n)$ holds that $bus(e_i) = bus(e_{i+1})$ and $component(e_{i+1}) = component(e_{i+2})$ for all $i = 1, 3, ..., n-2$ where $n > 2$. The function $start : \Pi \rightarrow E_{AG}$ denotes the first edge of a path. The function $end : \Pi \rightarrow E_{AG}$ denotes the last edge of a path. For example, for the architecture graph presented in Figure 3.4 the set of possible communication paths Π is:

$$\begin{aligned}\Pi = \{ \; & (e_1, e_3), (e_1, e_2, e_4, e_5), (e_1, e_2, e_4, e_6), (e_1, e_2), (e_1, e_3, e_5, e_4), \\ & (e_1, e_3, e_6, e_4), (e_3, e_2), (e_5, e_4), (e_6, e_4), (e_3, e_1), (e_5, e_4, e_2, e_1), \\ & (e_6, e_4, e_2, e_1), (e_2, e_1), (e_4, e_5, e_3, e_1), (e_4, e_6, e_3, e_1), (e_2, e_3), \\ & (e_4, e_5), (e_4, e_6) \; \}\end{aligned}$$

There exist two different component types, namely *basic components* C^B and *hierarchical components* C^H, $C = C^B \cup C^H$.

Definition 8 (Basic component C^B) *A basic component has no hierarchical refinement. All its interfaces are either type P or type R.*

A typical basic component is a microcontroller where the semiconductor manufacturer only provides information about the programming model of the device and the timing behavior of the interfaces. The exact internal structure of the microcontroller is not commonly known. Figure 3.5a) shows the graphical representation of the two basic architecture components. P-interfaces are drawn with a small filled box and R-interfaces are denoted with an empty box.

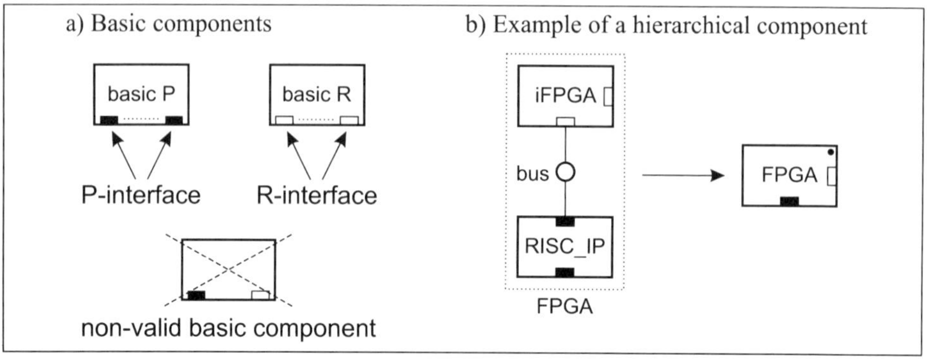

Figure 3.5: Component types.

If the internal structure of a component is sufficiently known, e.g., in case of custom ASIC design, hierarchical architecture components can be applied.

Definition 9 (Hierarchical component C^H) *A hierarchical architecture component abstracts from an internal architecture graph. This graph consists of connected basic and hierarchical components, as well as buses. Interfaces of hierarchical components are the unused interfaces of the internal components.*

Hierarchical components are marked with a black dot in their component symbol. Typical examples for hierarchical components include FPGAs with embedded processor cores for sequential task execution, partially reconfigured FPGAs, the Triscend A7 processor [Tri] with an ARM7 core and an additional configurable logic core. As an example, Figure 3.5b) shows a hierarchical component of an FPGA with an embedded RISC core. Such a component provides the concurrent execution of tasks written in VHDL (on internal component $iFPGA$) and the sequential execution of tasks written in C (on the internal component $RISC\_IP$) and may be an alternative to a discrete two device solution.

3.1.3 Mapping

A mapping specifies which component of the architecture graph will implement or execute which node and communication relation of the problem graph.

Definition 10 (Potential Bindings β^*) *The relation $\beta^* \subseteq (V \times C^B)$ denotes the set of potential bindings between nodes of the problem graph and basic components of the architecture graph.*

A single tuple $b = (v, c) \in \beta^*$ denotes an actual node implementation for a single target unit, e.g., a C-function for a processor, or a VHDL entity for an FPGA. Therefore, potential bindings β^* denote the set of all available node implementations. For example, in the context of hardware/software co-design each node has a hardware and a software implementation. Figure 3.6a) shows a set of potential bindings β^* (indicated by

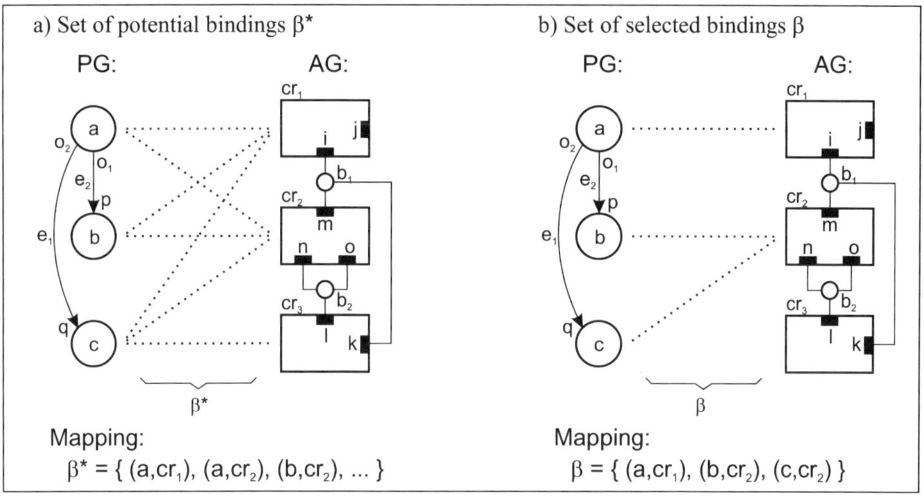

Figure 3.6: Mapping a problem graph to an architecture graph.

dotted lines) for the problem graph discussed in Figure 3.3 and the architecture graph discussed in Figure 3.4.

Based on the potential bindings β^*, dedicated bindings have to be selected for each node of the problem graph.

Definition 11 (Binding β) *The set $\beta \subseteq \beta^*$ selects bindings for each node of the problem graph. For a valid binding β holds $\forall v \in V : \exists(v, c) \in \beta$.*

Figure 3.6b) shows a set of selected bindings for the specification in Figure 3.6a).

3.2 Embedded System Model (*EPS*)

The EPS formalism is an extension of GPS and targets at the specification of synthesis problems for dataflow oriented embedded systems. These systems consist of connected general and special purpose processors, programmable as well as dedicated hardware components, and memories. The following list outlines the main differences between the formalisms EPS and GPS:

- The problem graph has two node classes: *executable nodes* to model a dedicated functionality, and *buffer nodes* for intermediate data. Furthermore, a problem graph has at least one dedicated node that models the node scheduling.

- An architecture graph has two component classes: *computing resource* that implements or executes nodes of the problem graph, and *storage* for data storage.

In terms of UML, EPS is a concrete model (non-cursive face in Figure 3.2). The remaining parts of this section introduce the EPS formalism and emphasize the differences between GPS and EPS.

3.2.1 Problem Graph

The EPS problem graph is a refined GPS problem graph and has two node classes:

- *executable* to model coarse-grained executable units, and

- *buffer* to model intermediate data storage.

Each node class provides two dedicated nodes types: *task* and *dispatcher* are executable nodes, *queue* and *register* are buffer nodes (see UML notation in Figure 3.7). An EPS node type is derived from the GPS node, inherits its properties, and adds characteristic features.

Definition 12 (Nodes of a PG) *The EPS model refines the nodes $V = V^T \cup V^D \cup V^Q \cup V^R$ of a problem graph into a set of tasks V^T, a set of dispatchers V^D, a set of queues V^Q, and a set of registers V^R.*

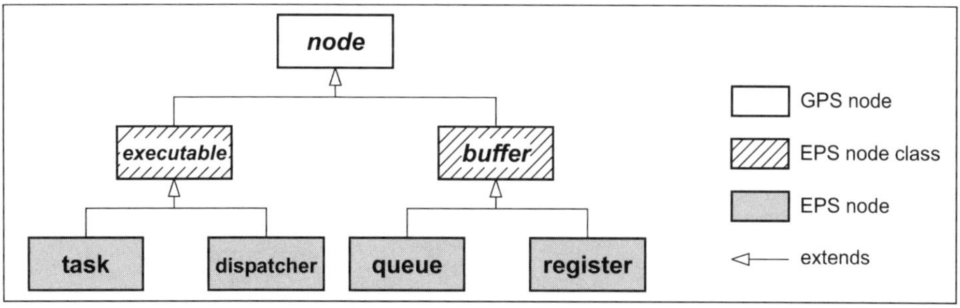

Figure 3.7: Node hierarchy of EPS problem graph nodes (UML).

In contrast to dataflow models like SDF, where a node is activated by the presence of data on all of its input ports, an EPS node is activated by sending a message to a dedicated control input port.

Definition 13 (Control input port i_c) *Each node $v \in V$ has a control input port $i_c \in I_v$.*

A control input port i_c is denoted by a black dot in the node's symbol. To interpret and react on data issued to the control input port i_c a node's implementation on a target unit involves a *node control finite state machine* (*ncFSM*). Nodes with unconnected control input ports are automatically activated and will be restarted upon their completion.

A *task node* $v \in V^T$ models a coarse-grained executable unit such as an FIR-filter in form of a VHDL entity or C function [ET98a]. It starts execution on receiving a *start* message on its control port i_c, executes its function which computes a set of outputs $o_1, ..., o_n$ from a set of inputs $i_1, ..., i_m$, and emits a *done* message at is control input port i_c on completion (see Figure 3.8a).

A *dispatcher node* $\sigma \in V^D$ models the activation of problem graph nodes. Its output ports O_σ are connected merely to control input ports. The problem-specific node activation sequence (schedule) is described by a *dispatcher finite state machine* (dFSM). By sending messages to connected nodes (dFSM actions) and receiving their reply messages (dFSM conditions) a dispatcher controls the execution of connected nodes.

A *queue node* $q \in V^Q$ models an intermediate buffer of limited size (see Figure 3.9a). It has destructive read and a non-destructive write access and provides one data input port and one data output port. The semantics of a queue (e.g., FIFO, LIFO) as well as the queue size is determined by the

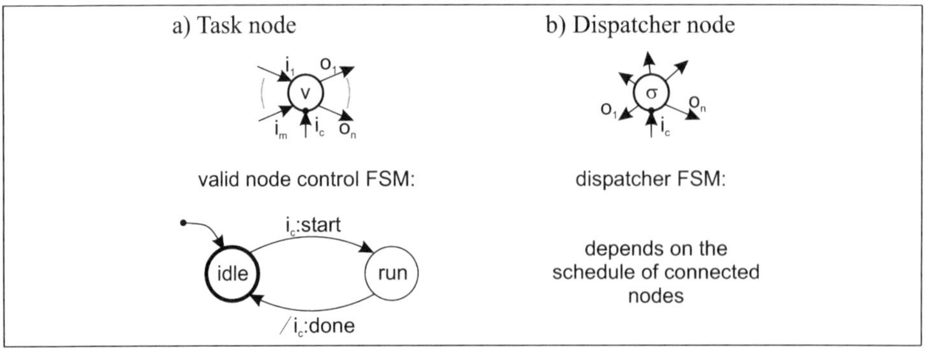

Figure 3.8: Task and dispatcher node.

implementation. Figure 3.9a) shows an example of a node control FSM for a queue. When the queue receives a *start* message on its control port i_c it enters the *run1* state. In this state, data items can be written into the buffer via port i and read via port o. When the buffer receives a *stop* command, it enters the *run2* state. In this state, data can still be written or read. The transition to the *idle* state is performed when the number of data items (tokens) equals a predefined value x (determined by the implementation). This property enables emitting a *done* message if the buffer contains the same number of tokens as in its initial state.

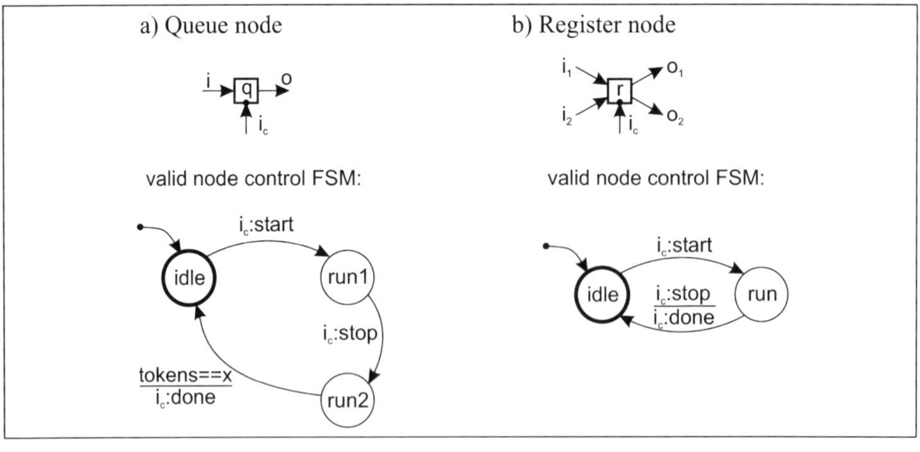

Figure 3.9: Buffer nodes.

A *register node* $r \in V^R$ is an intermediate buffer of limited size providing destructive write and non-destructive read access on an array of memory cells (see Figure 3.9b). The size as well as the addresses for reading/writing the register are defined by the implementation.

As an example, Figure 3.10a) shows a dispatcher σ whose outputs o_1 and o_2 are connected to the control input ports of a task v and a buffer b. In Figure 3.10b) a possible message sequence between the nodes is outlined where the dispatcher σ sends a *start* message to both nodes v and b. Some time later, v replies with a *done* message. Subsequently, the dispatcher stops the buffer operation. Figure 3.10c) shows the dFSM of the dispatcher σ causing the message sequence outlined in Figure 3.10b). The dFSM is automatically activated as the control input of the dispatcher is not connected.

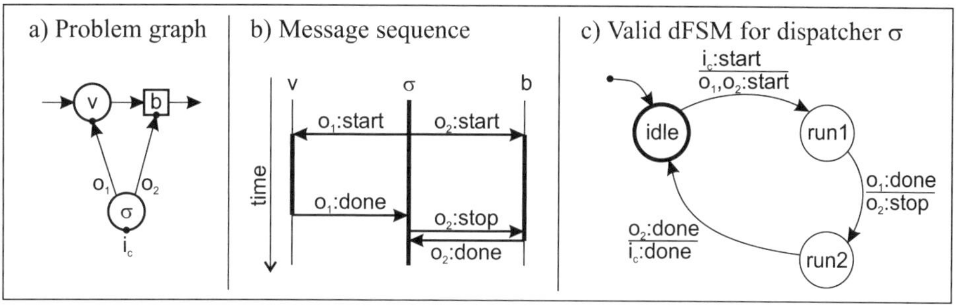

Figure 3.10: Specification of a simple system.

3.2.2 Architecture Graph

The EPS architecture graph is a refined GPS architecture graph that captures the target consisting of electronic units. There exists two component classes:

- *computing resource* that model units implementing or executing nodes of the problem graph, and
- *storage* to model temporary data storage.

The class computing resource provides two component types: *processor*, a component for sequential code execution, and *FPGA*, a reconfigurable component providing parallel and sequential code execution (as shown in Figure 3.11). The class storage has a single component type *memory*. Each

EPS component type is derived from the GPS component, inherits its properties, and adds characteristic features.

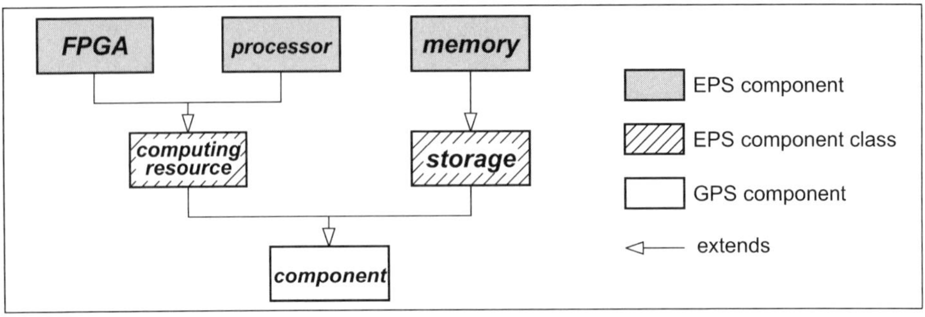

Figure 3.11: Hierarchy of EPS architecture graph components (UML).

Definition 14 (Components of an AG) *The EPS model refines the basic components $C^B = C^P \cup C^R \cup C^M$ of an architecture graph into a set of processor computing resources C^P, a set of reconfigurable computing resources C^R, and a set of memories C^M.*

A *processor computing resource* $c \in C^P$ models a microcontroller, processor, or DSP (Digital Signal Processor), and provides sequential code execution for problem graph nodes.

An *FPGA computing resource* $c \in C^R$ models a hardware programmable device, that provides parallel implementation for problem graph nodes. As an example, Figure 3.12 shows a model for the Triscend A7 [Tri] computing resource (block diagram shown in Figure 2.4). It is a hierarchical component that consists of two connected basic components: (i) *ARM7*, a model for the processor core, and (ii) *CSL*, a model for the configurable system logic matrix. *PIO* is a parallel I/O port provided by the CSL. *CSI* is the configurable system interconnect bus. *MI* is the interface to an external memory.

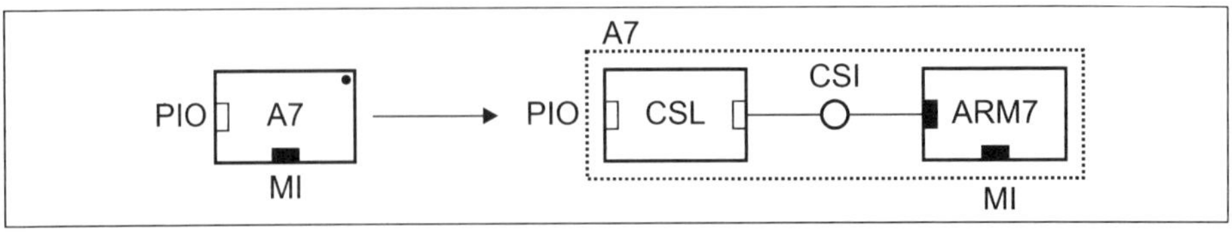

Figure 3.12: Hierarchical architecture model for the Triscend A7.

A *memory component* $m \in C^M$ models physical storage to hold data items of queue and register nodes. Additionally, a memory component enables to model memory-mapped data I/O, e.g., for sensor inputs. As an example, Figure 3.13 shows a simple EPS model where a Motorola *MC68340* microcontroller is connected to a XILINX *XCV1000* FPGA using a synchronous dual port SRAM *MCM69D536* (Motorola). A valid binding β specifies that tasks a, c, d as well as the dispatcher σ_1 are executed on *MC68340*, the intermediate queue q and the register r are implemented on *MCM69D536* and tasks b, e as well as dispatcher σ_2 are implemented on *XCV1000*.

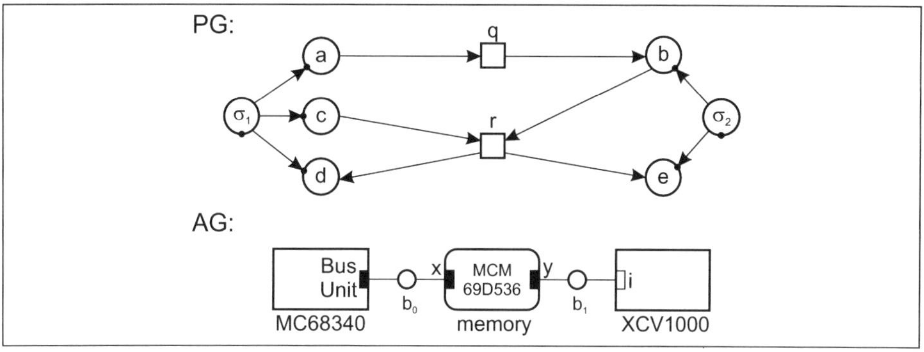

Figure 3.13: Problem specification using a memory component.

3.2.3 Mapping

An EPS mapping is quite similar to a GPS mapping. Additionally, regarding the later synthesis process, a communication path π has to be specified for each edge e of the problem graph. Obviously, such a path assignment is only valid if source and sink nodes of an edge e are bound to the corresponding start/end components of a selected path π.

Definition 15 (Edge binding τ) *The function $\tau : E_{PG} \rightarrow \Pi$ associates an edge with a communication path. τ assigns a valid path π to an edge e if the following constraints are satisfied:*

1. $(node(source(e)), component(intf(start(\pi)))) \in \beta$, *and*
2. $(node(sink(e)), component(intf(end(\pi)))) \in \beta$.

Furthermore, for the generation of interface circuits and device drivers for component interfaces of the architecture graph an interface generator has to be selected.

Definition 16 (Interface generator selection Λ) *The function $\Lambda : E_{PG} \rightarrow G$ selects an interface generator for an edge.*

A *feasible mapping* of a problem graph to an architecture graph in the context of EPS satisfies the following constraints:

C2 Each problem graph node is bound exactly once, i.e., $|\beta| = |V|$.

Comment: This constraint assures that a problem graph node is implemented or executed by one architecture component only. EPS does not support any kind of problem graph node migration between architecture components.

C3 For each edge e of the problem graph either

C3.1 the function τ is defined and the path $\pi = \tau(e)$ consists of two edges, or

C3.2 source and sink node of edge e are bound to the same component c, i.e.,

$$\{(node(source(e)), c), (node(sink(e)), c)\} \subseteq \beta.$$

Comment: The EPS formalism supports the specification of channels between adjacent components (constraint C3.1*) and on single components (constraint* 3.2*) only. However, by the introduction of (routing) tasks that just copy their input data to their output a long path can be broken down into a set of subsequent paths of length two (see Ex. 1).*

C4 Each computing resource c has at least one dispatcher node σ bound to it, i.e.,

$$\forall c \in (C^P \cup C^R) : \exists b = (\sigma, c) \in \beta : \sigma \in V^D.$$

Comment: Each computing resource executes or implements a set of problem graph nodes. These nodes have to be activated, which is the task of at least one dispatcher.

C5 Assume that a dispatcher σ is bound to a computing resource c.

C5.1 All tasks whose control inputs are connected to σ must be bound to computing resource c.

C5.2 If a dispatcher has been specified for a processor computing resource $c \in C^P$, it must be connected to the control inputs of all tasks of the computing resource c.

Comment: The implementation of a channel requires additional code (device drivers and interface circuitry). This introduces overhead concerning communication speed as well as program size. Subsequently, to minimize that overhead constraint 5.1 *has been defined. In case of processor computing resources, only one dispatcher is required (constraint* 5.2*).*

C6 Assume that $\pi = \tau(e)$ is the path selected for an edge e. The selected interface generator is $\Lambda(e) = g$. The ports of the involved problem graph nodes are $o = source(e)$ and $i = sink(e)$. It must hold that:

C6.1 $portProtocol(o) = protocol1(g)$

C6.2 $portProtocol(i) = protocol2(g)$.

Comment: Interface generators produce the necessary device drivers and interface circuitry to implement the required channels on the target platform. Constraints C6.1 *and* 6.2 *assure the compatibility between the protocols of the problem graph nodes and the protocols of the codes produced by the interface generators.*

As an example, consider the problem specification given in Figure 3.14. The problem graph consists of a set of tasks $a, ..., e$ whose control inputs are connected to a set of dispatchers $\sigma_1, ..., \sigma_3$. Note that a dispatcher's control input can be connected to another dispatcher. Here, σ_3 is connected to the control inputs of the dispatchers σ_1 and σ_2. This property enables hierarchical scheduling of tasks. For example, consider σ_3 as a general controller. In this case, the problem graph can be viewed as consisting of two subgraphs PG_1 and PG_2 containing tasks a, b, c and tasks d, e respectively. Each subgraph has its own dispatcher, i.e., σ_1 and σ_2. As an example for power optimization, each subgraph could be temporarily stopped from execution which is controlled by σ_3. However, the tasks have to support such an execution behavior. The architecture graph comprises two connected FPGA computing resources $fpga_1$ and $fpga_2$. Each problem graph node can be bound to any of the two computing resources. The edge

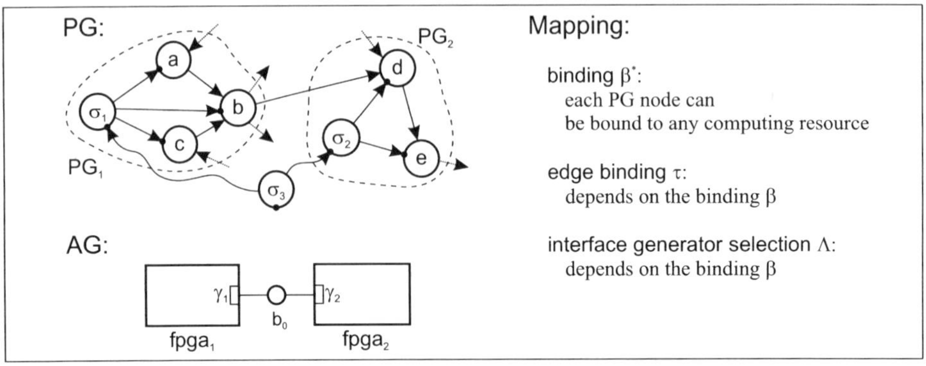

Figure 3.14: Dispatchers on FPGA computing resources.

binding τ as well as the selection of the interface generators Λ depends on the binding β.

3.2.4 Examples

The following examples present the specification of synthesis problems using the proposed EPS formalism.

Example 1 (Transducer) *A transducer is a hardware circuit providing protocol translation from its input port to its output port [Bor88]. Figure 3.15 shows the connection of two incompatible computing resources* cr_1 *and* cr_2*, i.e.,* $generators(\gamma_1) \cap generators(\gamma_4) = \varnothing$. *By using an in-*

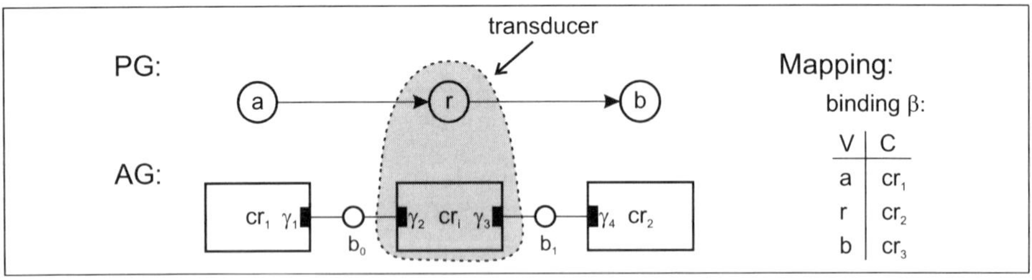

Figure 3.15: EPS model of a transducer.

termediate computing resource cr_i *the two components can be connected, if either (i)* $generators(\gamma_1) \cap generators(\gamma_2) \neq \varnothing$ *and* $generators(\gamma_3) \cap generators(\gamma_4) \neq \varnothing$*, or (ii)* cr_i *has R-type interfaces. To enable the communication between task* a *and task* b *bound to* cr_1 *and* cr_2 *respectively, a* *(routing) task r has been introduced into the problem graph that just copies its input to its output. The shaded area in Figure 3.15 consisting of the task r and the computing resource cr_i forms a model for a transducer.*

Example 2 (FunState: Basic Component) *Figure 3.16a) presents a basic FunState component (taken from Strehl [Str00]). The basic untimed FunState component consists of two parts, (i) a network N of connected storage units (e.g., $q_1 \ldots q_4$) and functions (e.g., $f_1 \ldots f_3$), and (ii) a finite state machine M.*

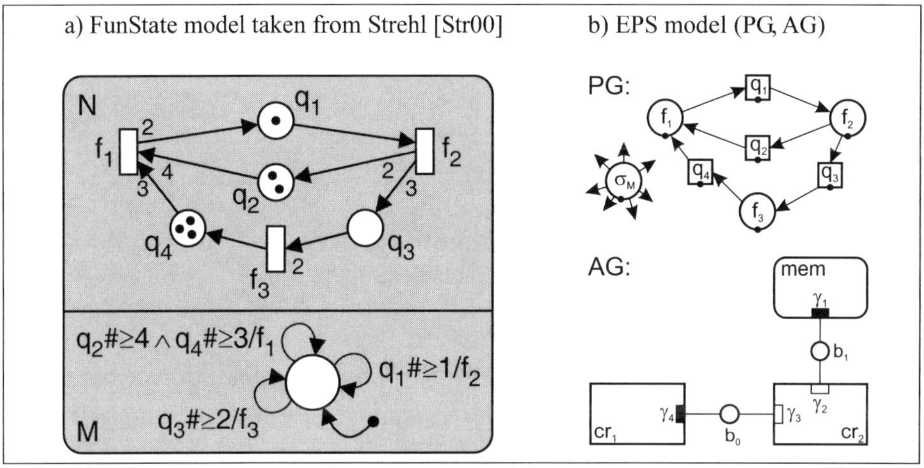

Figure 3.16: FunState: Basic component.

Valid problem and architecture graphs of a corresponding EPS specification are given in Figure 3.16b) where functions are modeled by tasks. A dispatcher σ_M whose output edges are connected to each control input port of the remaining nodes provides the state machine M of the FunState component. A mapping is not shown here.

Example 3 (Control System) *Figure 3.17a) shows a typical loop of a control system where a plant with an input vector $\overline{u}$ and an output vector $\overline{y}$ has to follow an input trajectory r. The shaded area denotes the control part of the system which (i) observes the plant by measuring the actual values of the system using sensors, (ii) compares them with the reference input r, and (iii) adjusts differences using its actors (signal $\overline{u}$).*

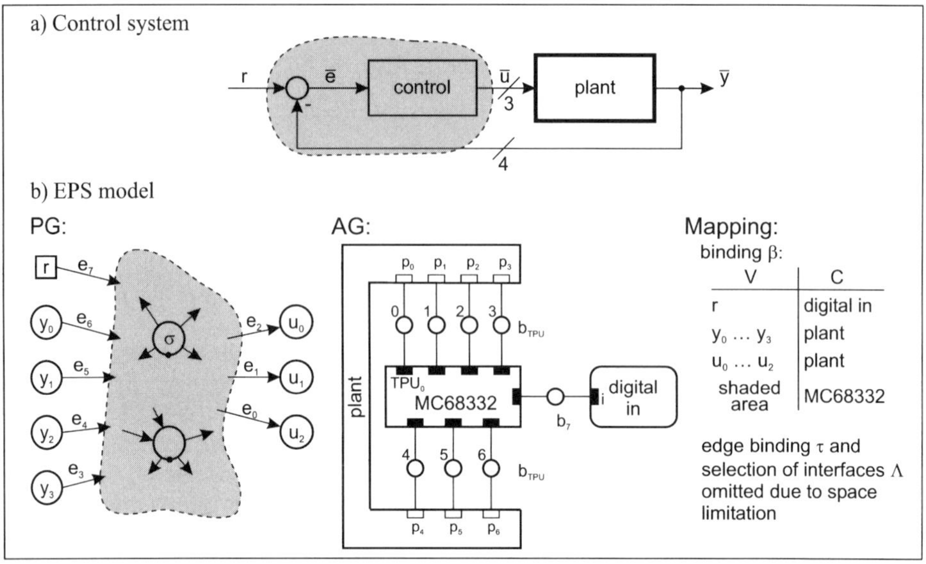

Figure 3.17: Control System.

The considered target architecture implementing the control part uses a Motorola MC68332 *microcontroller (see Figure 3.17b) which is well suited to cope with several (in-)dependent real-time I/O ports (time processing unit (TPU)). The environment (of the control system) is modeled by an FPGA computing resource* plant *representing the connections to the actual plant. Tasks modeling the control system's in- and outputs are bound to this resource. The required microcontroller's input and output ports are connected to the plant using bus nodes* $b_{TPU0}...b_{TPU6}$. *The memory mapped input* r *is modeled by the memory* digital in.

In the problem graph the shaded area indicates the behavior of the control part. The inputs and outputs are modeled via dummy tasks $y_0...y_3$ *and* $u_0...u_2$ *respectively. This is necessary to enable the generation of device drivers for the microcontroller's ports during the later synthesis phase. Note that only code for the microcontroller has to be generated which includes the shaded behavior part as well as appropriate input and output device drivers.*

Example 4 (MICROWIRE/PLUS Bus) *The COP8 microcontroller family of National Semiconductor Inc. provides solutions for small embedded system implementations. Currently, the family consists of about 30 microcontroller derivatives distinguishing in features like, e.g., on-chip RAM/ROM, number of I/O lines, etc. The series is added by a considerable range of peripheral devices such as EEPROMs, A/D-converters, and display drivers. All these devices support the proprietary serial 3(+1)-wire MICROWIRE/PLUS bus [SV95] providing simple component interconnection based on a master/slave communication scheme.*

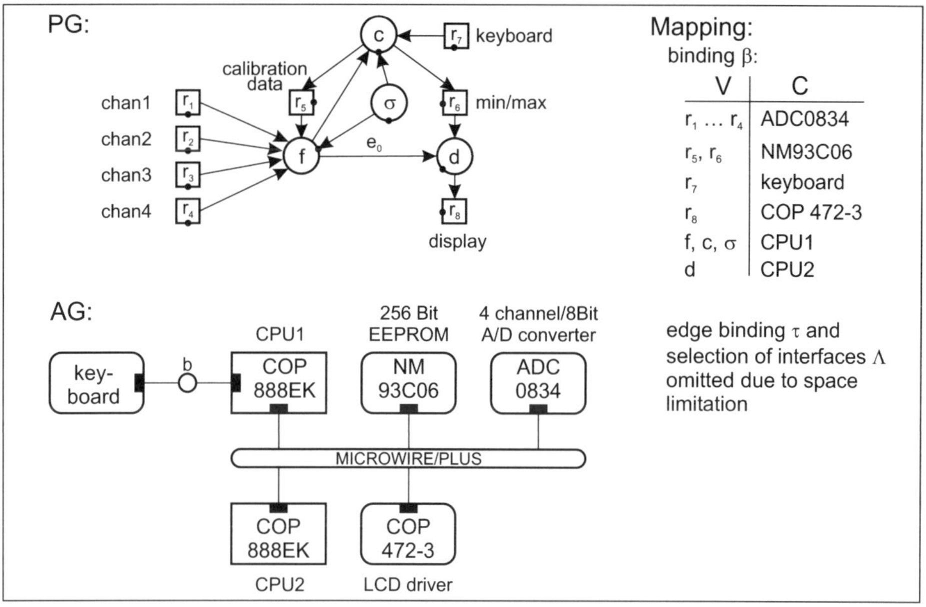

Figure 3.18: Simple measurement system.

Figure 3.18 shows the specification of a simple measurement system of three tasks f (filter), d (display), and c (control) using a typical bus architecture (MICROWIRE/PLUS). f reads from four analog inputs ($r_1 \ldots r_4$), filters and normalizes the data based on calibration data, and sends them to task d and task c respectively. Task d determines an overflow/underflow of the data values based on min/max values and writes the data to the display r_8. Using keyboard r_7 calibration data as well as min/max values can be defined.

3.3 Reconfigurable System Model (*RPS*)

Reconfigurable systems provide dynamic reconfigurable hardware computing resources to speed-up performance critical behavior parts [HW95]. Such resources may undergo a *repeated reprogramming during application execution* to optimize their workload. In the context of this thesis, these systems give rise to very interesting questions concerning the specification and implementation of the communication between dynamically reconfigured computing resources. To deal with reconfigurable systems the RPS formalism has been developed which is an extension of EPS. The main differences between the models RPS and EPS can be summarized as follows:

- The problem graph nodes are divided into groups. All nodes of a group are implemented by the same basic computing resource during a certain time period of application execution.
- The repeated reconfiguration of computing resources is modeled by the interplay of dedicated problem graph nodes. For that reason, the RPS formalism provides an additional node type.

In terms of UML, RPS is a concrete model (non-cursive face in Figure 3.2). The remaining parts of this section introduce the RPS formalism, emphasize differences between RPS and EPS, elaborate application scenarios, and outline related work.

3.3.1 Problem Graph

The dynamic reconfiguration of an FPGA computing resource requires that problem graph nodes bound to that resource have been divided into groups, called *configurations*. A single configuration δ denotes a set of problem graph nodes that are implemented by the *same basic computing resource during a certain time period.*

Definition 17 (Configuration Δ) *Δ denotes a set of all configurations $\delta \in \Delta$ of a problem graph.*

To specify embedded systems using dynamically reconfigured resources we use a *hierarchical specification approach* [EP00a] that consists of two layers:

- On a *static top layer* one or several "supervisory" nodes called *configurator(s)* supervise a set of dynamically reconfigured FPGA computing resources by downloading and starting their configurations.
- On a *dynamic bottom layer* a dispatcher node is assigned to each configuration. A dispatcher's outputs are connected to the control inputs of the remaining nodes of the same configuration. Its own control input is connected to one supervising configurator node. The dispatcher activates the connected nodes and determines the end of a configuration.

As an example, Figure 3.19 sketches the hierarchical specification of the reconfiguration for a problem graph. The problem graph nodes (not shown for simplification) have been divided into a set of configurations $\delta_1, ..., \delta_5$ (indicated by dashed lines). Each configuration δ_i has its own dispatcher σ_i whose control input is connected to the supervisory configurator ρ. The binding is indicated by rectangular boxes. Hence, the configurations will be implemented on either $fpga_1$ or $fpga_2$. The configurator ρ will be implemented on the computing resource $host$.

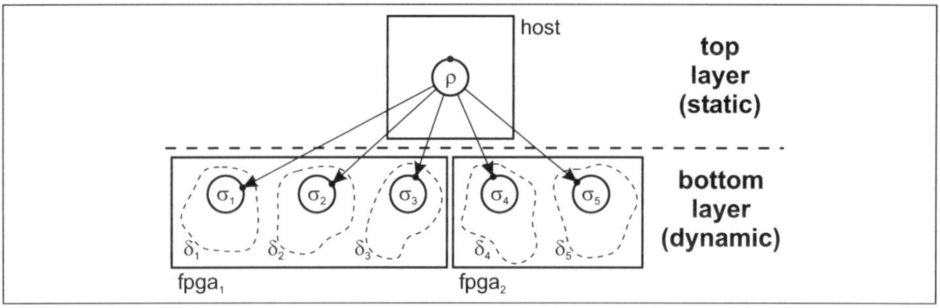

Figure 3.19: Hierarchical specification of an FPGA reconfiguration.

In general, an RPS problem graph includes the same specification features and node types as an EPS problem graph but additionally has a node type *configurator* (see UML [PJ99] notation in Figure 3.20).

Definition 18 (Nodes of a PG) *The RPS model refines the nodes* $V = V^T \cup V^D \cup V^C \cup V^Q \cup V^R$ *of a problem graph into a set of tasks* V^T, *a set of dispatchers* V^D, *a set of configurators* V^C, *a set of queues* V^Q, *and a set of registers* V^R.

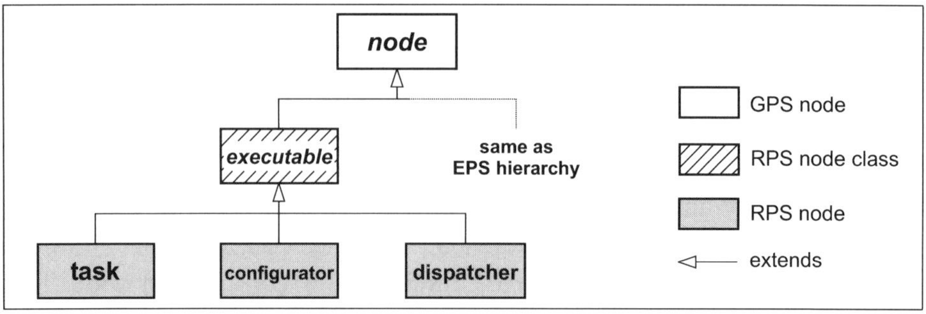

Figure 3.20: Node hierarchy of RPS problem graph nodes (UML).

A *configurator* $\rho \in V^C$ is a problem graph node that (i) models configuration switches of dynamically reconfigured computing resources, and (ii) is able to initiate downloading and starting of FPGA configurations. Its output ports O_ρ are connected to control input ports of dispatcher nodes. A configurator implements a *configurator finite state machine* (*cFSM*) that describes a run-time configuration sequence. By sending messages to connected dispatchers (cFSM actions) and receiving their reply messages (cFSM conditions), a configurator supervises a set of run-time reconfigured FPGA resources. Each problem graph with run-time reconfigured components requires at least one configurator node.

The two suggested layers *top* and *bottom* jointly implement an overall schedule of problem graph nodes and configurations by executing the dispatcher (dFSM) and configurator (cFSM) state machines. The schedule is either statically determined at design time or dynamically by the dispatchers and configurators at run-time (not subject of the thesis). A schedule is *static* if the sequence of node/configuration executions does not depend on run-time conditions and has been fixed during compile-time. A schedule is *dynamic* if the sequence of node/configuration executions depends on run-time conditions, i.e., the schedule is influenced by reply messages of nodes and configurations. In any case, three sources of constraints can be identified for the overall scheduling of nodes and configurations: (i) the schedule has to reflect the read/write order of ports of problem graph nodes, (ii) the schedule must resolve non-determinism in the specification either at compile-time or at run-time using dispatchers and configurators, and (iii) the schedule must respect configuration borders given by the partitioning into configurations. Scheduling is a non-trivial task and it has been shown [DJ98] that careless assignment of problem graph nodes to configurations may lead to infeasible schedules.

3.3.2 Architecture Graph

The architecture graph of an RPS specification comprises the same components as an EPS specification. However, if an architecture graph includes dynamic reconfigured computing resources, configurator nodes appear in the problem graph that are bound to *host* computing resources.

Definition 19 (Host computing resource H) *The function $H : ((C^P \cup C^R) \times C^R) \rightarrow \{0, 1\}$ evaluates to 1, i.e., $H(cr_1, cr_2) = 1$, if $cr_1 \in (C^P \cup C^R)$ is a host computing resource of computing resource $cr_2 \in C^R$.*

The property of being a host computing resource cr_1 for a certain reconfigurable computing resource cr_2 expresses the fact that the host cr_1 is able to (i) (re)configure the resource cr_2, and (ii) start and supervise the dispatchers of the configurations on cr_2.

3.3.3 Mapping

A feasible mapping in the context of an RPS specification includes the mapping features of an EPS specification. Additionally, four steps are necessary. Essentially, it includes the definition of configurations, the assignment of nodes to configurations, the insertion of dispatcher and configurator nodes (for the specification of the reconfiguration) resulting in an extended problem graph PG', as well as the consideration of a set of constraints. This section presents a simple problem as well as the proposed steps to describe an appropriate mapping.

Step 1: *Define potential assignments to configurations and components*

The problem graph nodes V have to be associated with configurations Δ and configurations Δ have to be associated with architecture components C^R.

Definition 20 (Potential Node-Configuration Assignments Φ^*) *Φ^* is a relation $\Phi^* \subseteq (V \times \Delta)$ denoting the set of potential assignments of problem graph nodes to configurations.*

A single tuple $\phi = (v, \delta) \in \Phi^*$ denotes the potential membership of a node v to a configuration δ.

Definition 21 (Potential Configuration-Component Assignments Ψ^*) *Ψ^* is a relation $\Psi^* \subseteq (\Delta \times C^R)$ between configurations and reconfigurable computing resources of the architecture graph.*

A single tuple $\psi = (\delta, c) \in \Psi^*$ denotes the potential membership of a configuration δ to a component c. Figure 3.21 shows a set of potential assignments Φ^* and Ψ^* for a given problem and architecture graph. For simplification, the potential binding β^* is given in textual representation.

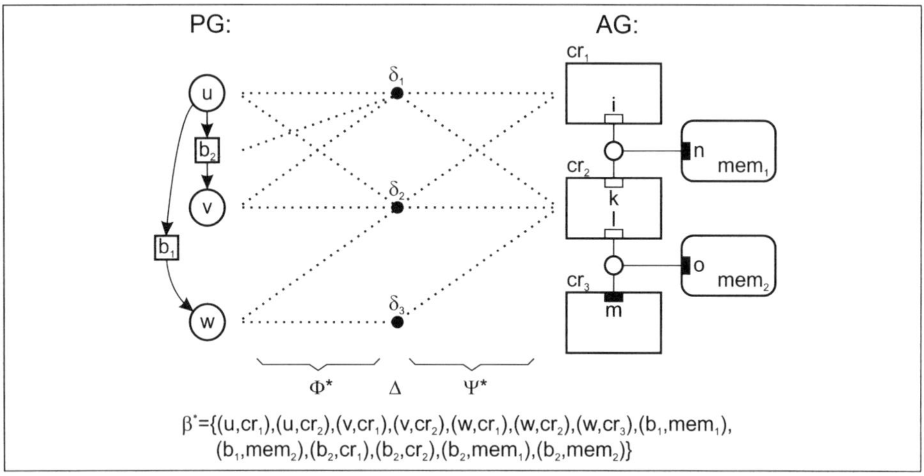

Figure 3.21: Potential assignments Φ^* and Ψ^*.

Step 2: *Select appropriate assignments*
Based on the potential assignments Φ^* and Ψ^* dedicated assignments Φ and Ψ have to be selected for later synthesis.

Definition 22 (Node-Configuration Assignment Φ) *The set $\Phi \subseteq \Phi^*$ selects node-configuration assignments from the set of possible assignments.*

Definition 23 (Configuration-Component Assignment Ψ) *The set $\Psi \subseteq \Psi^*$ selects configuration-component assignments from the set of possible assignments.*

Step 3: *Insert the hierarchical reconfiguration structure*
Based on the assignments Φ and Ψ, the necessary hierarchical reconfiguration structure has to be added into the problem graph PG.

3.1 (*Bottom Layer*) For each used configuration, an additional dispatcher node has to be inserted into the problem graph and bound to the corresponding FPGA computing resource. Its output ports have to be connected with control input ports of nodes assigned to the same configuration.

3.2 (*Top Layer*) At least one configurator node has to be inserted into the problem graph and bound to a host computing resource. Its output ports have to be connected with control input ports of the dispatchers of the configuration (see step 3.1).

The adding of dispatcher and configurator nodes to the problem graph PG results in an extended problem graph PG'. Subsequently, these additional nodes have to be bound to appropriate computing resources leading to a binding β'. Additionally, the inserted dispatchers have to be assigned to the corresponding configurations resulting in a node-configuration assignment Φ'. As an example, Figure 3.22 shows the refined specification after applying step 2 and 3 to the problem provided by Figure 3.21. The computing resource cr_2 has three configurations. For example, configuration δ_1 comprises the task node u and the dispatcher node σ_1. The configurator ρ is bound to the computing resource cr_3 and is connected to the control inputs of the three dispatchers (only indicated for simplification).

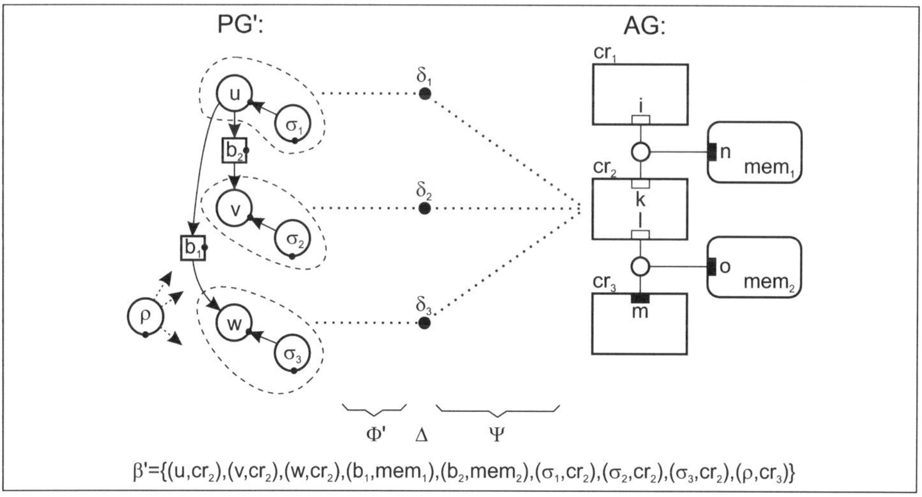

Figure 3.22: Refined RPS specification after applying step 2 and 3.

Step 4: *Consider constraints*
A feasible mapping in the context of RPS must satisfy a set of constraints. Besides the constraints *C3*, *C4*, *C5*, and *C6* of the feasible mapping of an EPS model a feasible mapping of an RPS model includes the following constraints:

C7 Assume that a configurator ρ is bound to a basic computing resource cr_1.

- **C7.1** Assume that the configurator ρ is connected to a dispatcher which is bound to a computing resource cr_2. In that case function $H(cr_1, cr_2)$ must evaluate to 1.
- **C7.2** All dispatchers whose control inputs are connected to ρ must be bound to a basic computing resources other than cr_1.
- **C7.3** A configurator supervises either all or none configurations of a certain computing resource.

Comment: Constraint C7.1 *assures that the configurator is bound to an appropriate host computing resource of each supervised FPGA. Constraint* C7.2 *prevents that a configurator reconfigures the FPGA that implements the configurator itself. Constraint* C7.3 *expresses the fact that an FPGA may have only one supervising configurator. Neglecting constraint* C7.3, *configurators supervising the same computing resource had to communicate to determine the next valid configuration.*

C8 Assume that there exists a configuration $\delta \in \Delta$ such that there is a tuple $(v, \delta) \in \Phi$ and a tuple $(\delta, c) \in \Psi$. In that case, there has to exist a corresponding binding, i.e.,

$$\exists \delta \text{ such that } (v, \delta) \in \phi \text{ and } (\delta, c) \in \psi \longrightarrow (v, c) \in \beta.$$

Comment: This constraint assures that configurations of an FPGA contain only node implementations that are actually bound to the FPGA.

C9 Assume that an edge e connects two nodes v and w, respectively. Both nodes are bound to the same reconfigurable basic computing resource. In that case, both nodes have to be assigned to the same configuration $\delta \in \Delta$, i.e.,

$$\{(v, \delta), (w, \delta)\} \subseteq \Phi.$$

Comment: This constraint considers the fact that nodes can not communicate between (exclusive) configurations of the same basic computing resource.

3.3.4 Application Scenario 1: Simple Model

In a first scenario of an RPS specification each problem graph node is bound exactly once, i.e., constraint *C2* of the EPS formalism is considered. Furthermore, for each problem graph node n there exists at most one tuple $(n, \delta) \in \Phi$. For each configuration δ there exists at most one tuple $(\delta, c) \in \Psi$. As an example, Figure 3.23 shows two assignments of Φ, Ψ, and corresponding bindings β for the problem suggested in Figure 3.21. For simplification, problem graph nodes as well as architecture components are merely indicated by black dots. In Figure 3.23a) computing resource cr_1 has one configuration implementing buffer b_2; computing resource cr_2 implements two configurations δ_2 and δ_3. In Figure 3.23b)

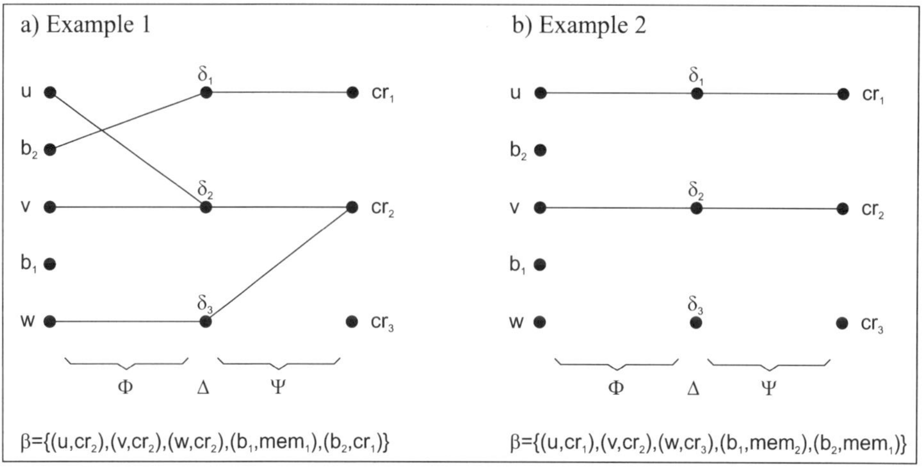

Figure 3.23: Scenario 1: Examples of Φ, Δ, and Ψ.

node w is not assigned to a configuration but is just bound to computing resource cr_3.

As further example, an EPS model is considered that has been extended to an RPS model. Figure 3.24a) presents the EPS model. The tasks are bound to the computing resource *fpga* and buffers q_1 and q_2 are bound to

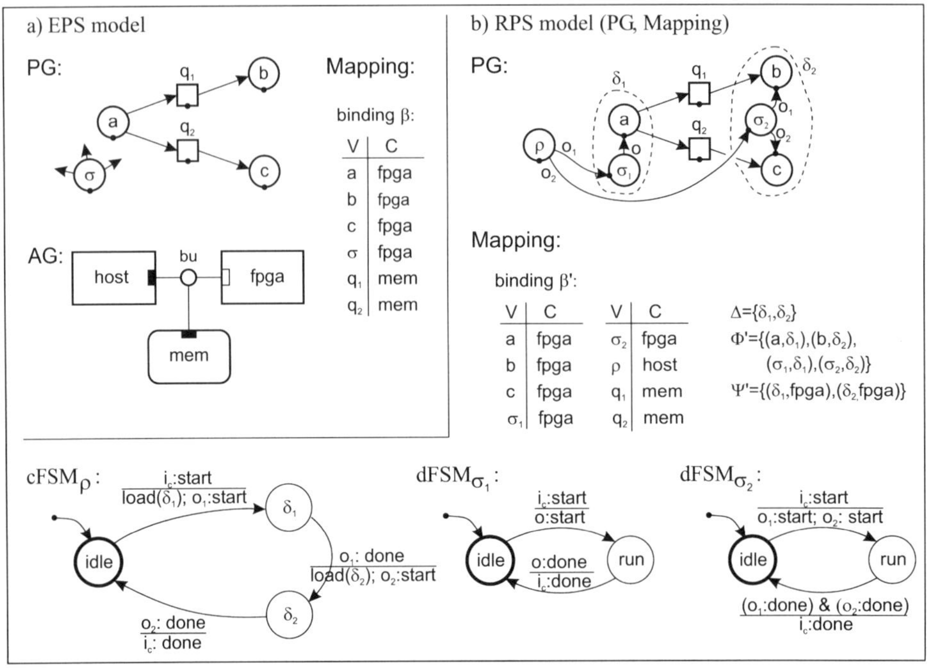

Figure 3.24: Simple example for application scenario 1.

the memory mem. The dispatcher's outputs are connected to the control inputs of the three tasks a, b, and c (not shown here for simplification). Now, it is assumed that the *fpga* has two configurations (see dashed areas in Figure 3.24b). For each of these configurations a dispatcher has been introduced that supervises the nodes of the configuration. For example, the dispatcher σ_2 is connected to the control inputs of the problem graph nodes b and c. To supervise the two configurations a configurator node ρ has been inserted. It is connected to the control inputs of the dispatchers σ_1 and σ_2. The configurator itself is bound to the host.

The schedule of the two configurations and nodes is determined by the interplay between $cFSM_\rho$, $dFSM_{\sigma_1}$, and $dFSM_{\sigma_2}$ respectively. Figure 3.24b) shows examples of appropriate FSMs. To reconfigure the *fpga* the configurator ρ executes a function $load()$ that (i) reads *fpga* programming data from a memory, and (ii) writes it to the *fpga* programming input (not shown here). Note that careful scheduling analysis is required to ensure liveliness of the system. For example, starting the configuration sequence by downloading configuration δ_2 causes a deadlock as nodes b and c will wait forever to get their data from the queues q_1 and q_2.

3.3.5 Application Scenario 2: Sharing Tasks

This scenario considers systems where a single task in the problem graph is associated with several *exclusive* configurations either on the *same* or *adjacent* computing resources. Such tasks are called *shared tasks*. They are the base for systems that actually remove a task's implementation from an FPGA, and at a later point in time, download and (re)start it again as part of the same or another configuration. As an example, Figure 3.25a) shows the sharing of task u between the configurations δ_1 and δ_2. Both configurations are associated with the same computing resource cr_2. Figure 3.25b) outlines a specification where task u is shared between configurations δ_1 and δ_2 that are associated with adjacent computing resources cr_1 and cr_2.

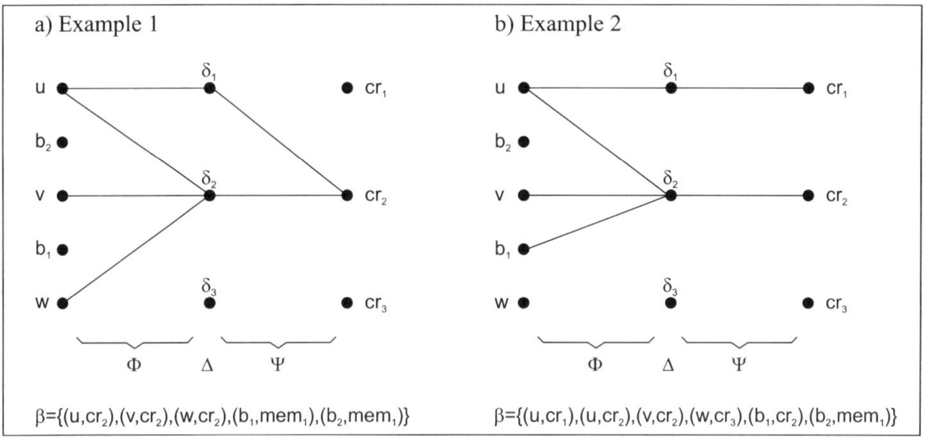

Figure 3.25: Scenario 2: Examples of Φ, Δ, and Ψ.

Essentially, to share a node n between a set of configurations $\delta_1 ... \delta_m$ constraints *C3* ... *C9* as well as the following constraints have to be considered:

C10 The configurations $\delta_1 ... \delta_m$ have to be exclusive.

Comment: Node n is specified once in the problem graph. Subsequently, an implementation of the problem graph may have at most one implementation of node n at the same time.

C11 Assume that S is the set of problem graph nodes that are connected to node n via a single edge. The nodes in S as well as node n are bound to the same computing resource (binding β). In that case, all nodes in S have to be associated with configurations $\delta_1 ... \delta_m$.

Comment: This constraint is quite similar to constraint C9. *In each configuration a shared task has to be connected to the same problem graph nodes to assure the specified behavior.*

As an example, consider the RPS specification outlined in Figure 3.26a) where three tasks x, A, and y communicate via buffers b_1 and b_2. Assume that x and y have considerably large implementation whereas A is rather small. To avoid the usage of a large FPGA in terms of area, the configurations δ_1 and δ_2 have been introduced. Furthermore, by sharing the implementation of task A between the configurations the overall execution time can be reduced. Note that the control input of node A is connected to the dispatchers of both configurations. However, due to the exclusive use of the configurations it actually never happens that both dispatchers are connected concurrently.

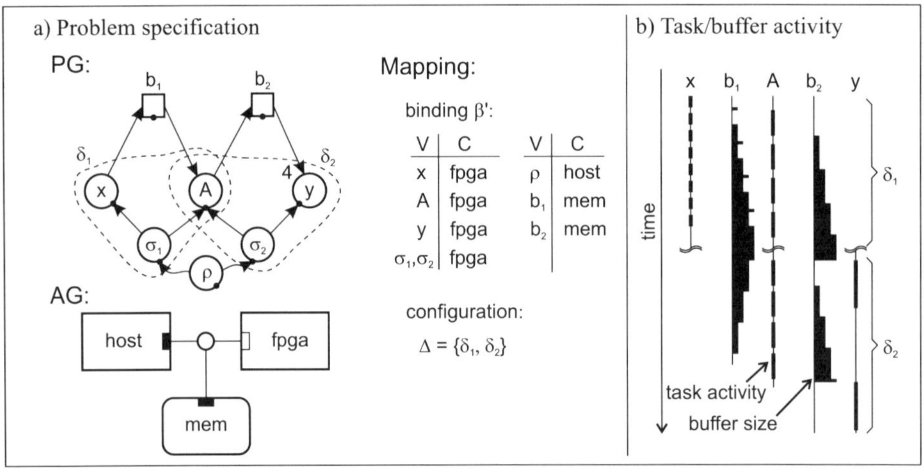

Figure 3.26: Sharing tasks between configurations on the same FPGA.

Figure 3.26b) shows the task and buffer activity during a certain time period. As soon as task x has finished and has written its first data value into the buffer b_1 task A can start its execution. Subsequently, x and A execute concurrently as long as x reaches an end of its repeated execution. After awaiting the end of A's current execution the configuration is switched to δ_2. As soon as δ_2 is on the $fpga$ computing resource, A and y start their execution. Therefore, A operates on the remaining data values in

b_1 and y at first uses the values produced by A in the previous configuration δ_1. After A's end of execution y starts its last iteration.

As a further example, consider the specification in Figure 3.27a) which is similar to Figure 3.26. Here, the target architecture consists of two adjacent FPGAs, a host computing resource as well as a memory. Task A is

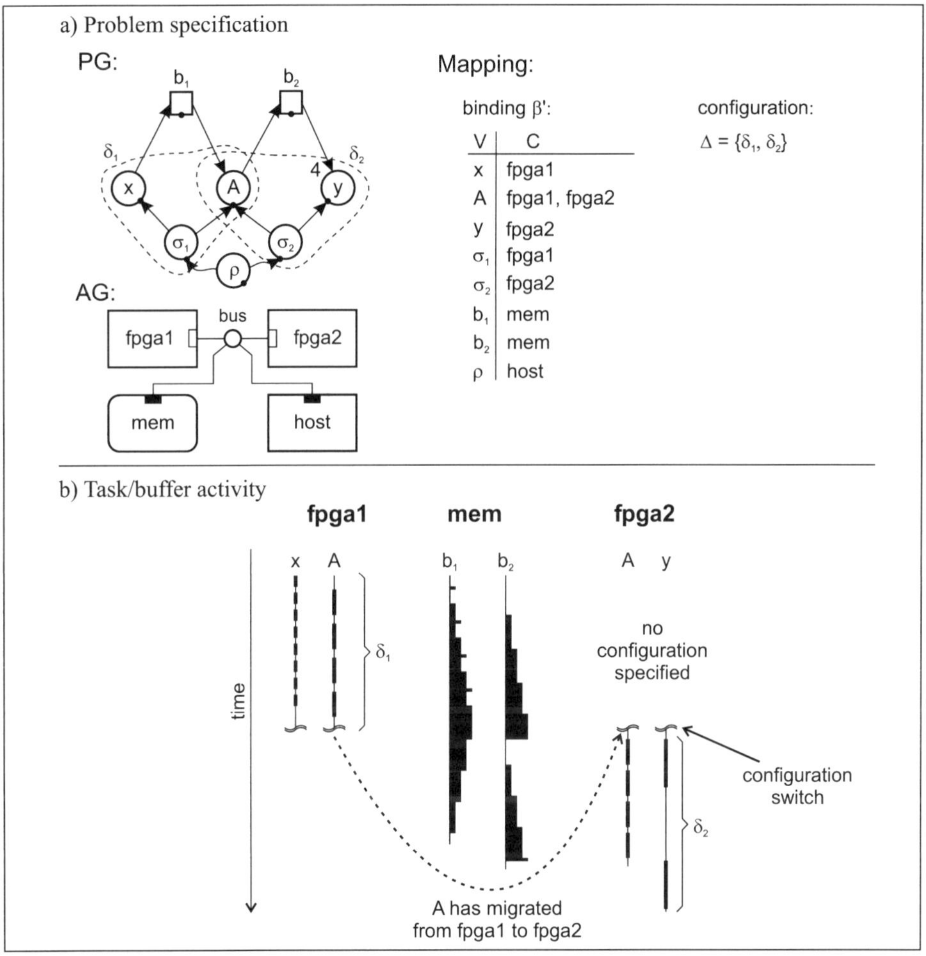

Figure 3.27: Sharing tasks between configurations of adjacent FPGAs.

now part of two exclusive configurations that are implemented on adjacent FPGAs. Figure 3.27b) outlines task and buffer activity over a certain time period for both FPGAs and the memory. Note that the configurations δ_1 and δ_2 have to be exclusive to prevent malfunction. For this reason, *fpga2 may not* be configured with configuration δ_2 as long as configuration δ_1 is being implemented on *fpga1*. As shown in Figure 3.27b) by switching the exclusive configuration δ_1 on *fpga1* to configuration δ_2 on *fpga2*, task A has actually migrated.

3.3.6 Application Scenario 3: Suspendable Tasks

The last section showed that a task can be migrated between FPGAs if its ncFSM is in the idle state. But, in the sense of system optimization, it can be advantageous to just suspend a task's execution, possibly migrate the task, and resume it at a later point in time. This section introduces tasks that provide suspending/resuming on FPGA resources.

The main issue of a corresponding task implementation is to provide a possibility to save/restore the *context*. The context comprises the state of the ncFSM and presumably states of the task's core FSMs, registers and memories to assure a proper task restart.

In the following, it is assumed that a context has been defined and the focus is on the appropriate specification. Figure 3.28a) shows the problem graph representation of a suspendable task v (one data input port i_1 and one data output o_1). It consists of the task v itself and a queue b_c to save

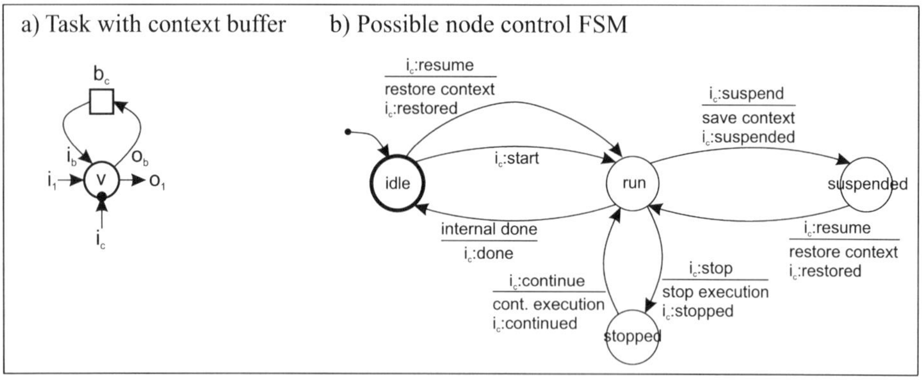

Figure 3.28: Suspendable task.

the task's context via output port o_b and restore it via input port i_b. Figure 3.28b) outlines a potential ncFSM consisting of four states that include a temporarily stop of the task's execution:

- *idle* is the initial state after downloading. On receiving a *start* or *resume* message on its control input either (i) the ncFSM immediately enters the state *run*, or (ii) the context stored in queue b_c is restored and then the state *run* is entered.

- In the *run* state the task executes its function. On completion the ncFSM reenters the *idle* state. On receiving a *stop* message on its control input the task stops its execution and enters the *stopped* state. If it gets a *suspend* message the task saves its context in queue b_c and enters the state *suspended*.

- In the *suspended* state the task can be removed from the FPGA computing resource by a configuration switch. If the task is not removed it can be resumed by sending a *resume* message to the control input. In this case, the context is restored and the *run* state is reentered.

- In the *stopped* state the task has stopped its execution and is not able to communicate via its ports apart from the control input. Hence, the clock on the task's core can be disabled to reduce the dynamic power consumption. If a *continue* message is sent to the control input the task awakes again and continues its execution.

To assure that a stored context does not get lost the following constraint has to be considered:

C12 Assume that a suspendable task is bound to an FPGA computing resource. A buffer that stores the context of the task can be bound to an architecture component c if at least one of the following cases arise:

C12.1 c is an adjacent memory.

C12.2 c is an adjacent basic FPGA computing resource that may not be reconfigured as long as the context is stored in the buffer.

Comment: This constraints assures that the buffer content remains fixed as long as the suspendable task is not resumed.

As an example, consider the problem specification in Figure 3.29a) where a task A is shared between two configurations δ_1 and δ_2. Buffer b_c is able to store the context of A. The architecture graph consists of a $host$, an FPGA computing resource *fpga*, and a memory mem. Figure 3.29b) shows a valid mapping where the $fpga$ implements a set of three configurations and the buffers are bound to mem. The diagram shows a valid configuration sequence for computing resource *fpga* as well as task and buffer activity. During the active time of configuration δ_1 task A is forced by σ_1 to save its context to b_c and enter its suspended state. Afterwards, the configuration is switched to δ_3. Finally, δ_2 is downloaded and A restores its context as forced by σ_2 and continues its execution.

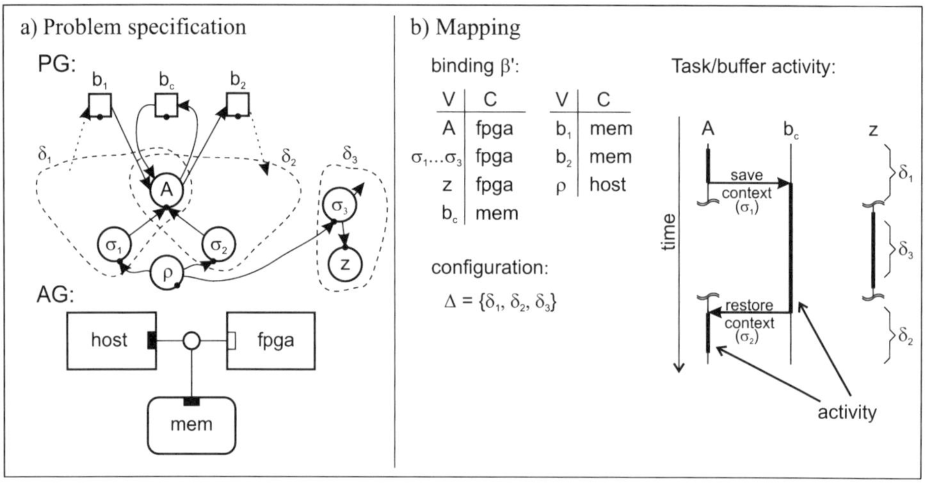

Figure 3.29: Scenario 3: Suspendable task.

Brief summary about suspendable tasks

The main advantages of suspendable tasks on FPGAs are (i) area optimization as several different "ready" tasks can co-exist in different configurations that are repeatedly downloaded, (re)started and suspended (OS approach), (ii) power optimization as a suspended task does not consume dynamic power, and (iii) task migration between configurations during execution.

These advantages are paid with (i) an increased area overhead as an appropriate ncFSM definitely has more states as an ncFSM not supporting this feature, (ii) an extended task execution time as saving/restoring a context takes time, and finally (iii) memory area to store the context.

3.3.7 Application Scenario 4: Virtual Configurations

This scenario considers systems where single configurations can be replaced on an FPGA by "pin-compatible" configurations. Such interchangeable configurations are called *virtual configurations*. Basically, virtual configurations use the same interface circuitry for communication but can have different associated problem graph nodes. Potential applications include: (i) configurations implementing the same behavior but differentiating in features like power consumption and execution speed, and (ii) configurations that are loaded over a network [LNTT01] to provide a system upgrade.

Definition 24 (Virtual Configurations Θ) *The function* $\Theta : \Delta \times \Delta \longrightarrow \{0, 1\}$ *evaluates to* 1, *if the implementations of two configurations can replace each other in a "pin-compatible" sense.*

As an explanatory example, consider the problem specification in Figure 3.30 which is quite similar to Figure 3.22. Here, task x as well as

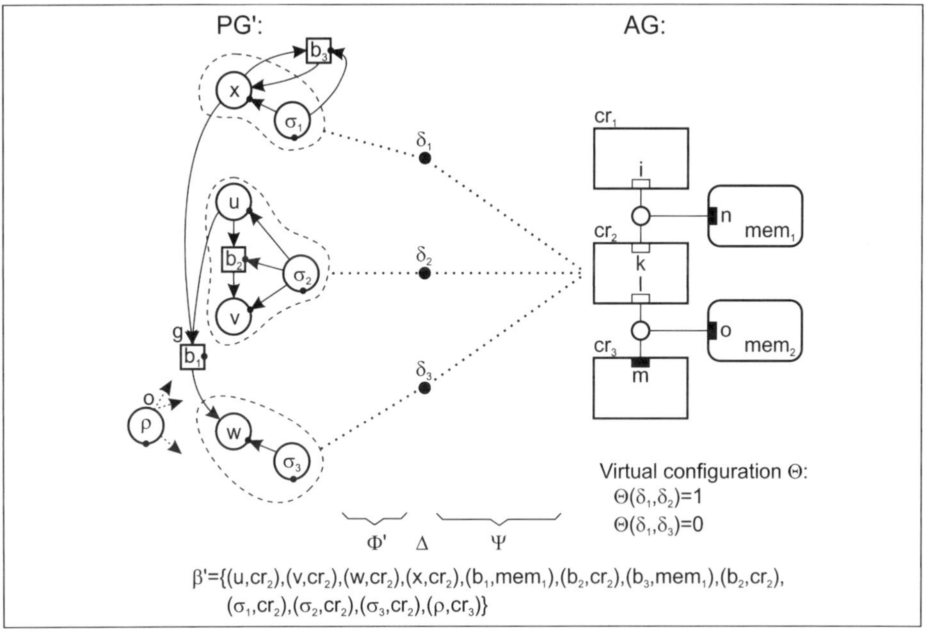

Figure 3.30: Scenario 4: Virtual configurations.

buffer b_3 have been added. Task x is connected to the same input port g of buffer b_1 as task u. The control input of the dispatchers σ_1 and σ_2 are connected to the same output port o of the configurator ρ. Buffer b_3 is only connected to nodes of configuration δ_1 and acts as temporal buffer. In this situation, configurations δ_1 and δ_2 are virtual configurations and can replace each other. Quite obviously, as δ_1 and δ_2 are interchangeable they have to be used exclusively in a schedule. Hence, a valid configuration sequence is $\delta_1; \delta_3; \delta_2; \delta_3; \delta_2; ....$

Basically, virtual configurations $\delta_1 ... \delta_n$ that are replacable among themselves consider constraints *C2* ... *C10* as well as the following constraints:

C13 Each configuration $\delta_1 ... \delta_n$ is merely assigned to the same basic architecture component, i.e., $\forall \delta \in \{\delta_1; ...; \delta_n\} : (\delta; c) \in \Psi$.

Comment: Configuration replacement is only possible on the base of single basic architecture components.

C14 Assume that two edges e_1 and e_2 are connected to the same port.

C14.1 Their edge binding τ must be the same, i.e., $\tau(e_1) = \tau(e_2)$.

C14.2 The port protocols of the nodes that are connected by the edges e_1 and e_2 have to be the same, i.e., $portProtocol(source(e_1)) = portProtocol(source(e_2))$ and $portProtocol(sink(e_1)) = portProtocol(sink(e_2))$.

Comment: Corresponding edge implementations have to use the same communication path (constraint C14.1*). Furthermore, to use copies of interface circuitry in each configuration the port protocols of the nodes must be the same (constraint* C14.2*).*

C15 A buffer can act as temporal memory for nodes of a configuration δ if the buffer

C15.1 is not assigned to configuration δ, and

C15.2 is connected only to nodes of configuration δ.

Comment: The content of such a buffer contains useful data for a single configuration only. As soon as a configuration switch arises the content becomes invalid.

As an example, buffer b_3 in Figure 3.30 is a such a temporal memory.

C16 Assume that E^{δ_1} denotes the set of edges that connect nodes assigned to configuration δ_1 with nodes not assigned to δ_1. Edges connecting buffers in the sense of constraint *C15* are not included in E^{δ_1}. Assume that $\mathcal{P}$ denotes the set of node ports that have been connected by the edges in E^{δ_1} and belong to nodes not assigned to configuration δ_1.

C16.1 Each configuration $\delta_2, ..., \delta_n$ may be connected to ports in $\mathcal{P}$ only.

C16.2 It must hold that $|E^{\delta_1}| = ... = |E^{\delta_n}| = |\mathcal{P}|$.

Comment: This constraint assures that each virtual configuration $\delta_1, ..., \delta_n$ merely uses copies of the same interface circuitry.

As an example, in Figure 3.30 the set of common port is $\mathcal{P} = \{o, g\}$.

3.3.8 Examples

This section outlines two examples specifying reconfigurable systems.

Example 5 (Continuous data stream) *Consider a system with a reconfigurable computing resource cr (see architecture graph AG in Figure 3.31). Assume that cr has to route a continuous data stream from its interface γ_x to its interface γ_y. Such a requirement is modeled by using a task r in the problem graph that just copies its input data to its output.*

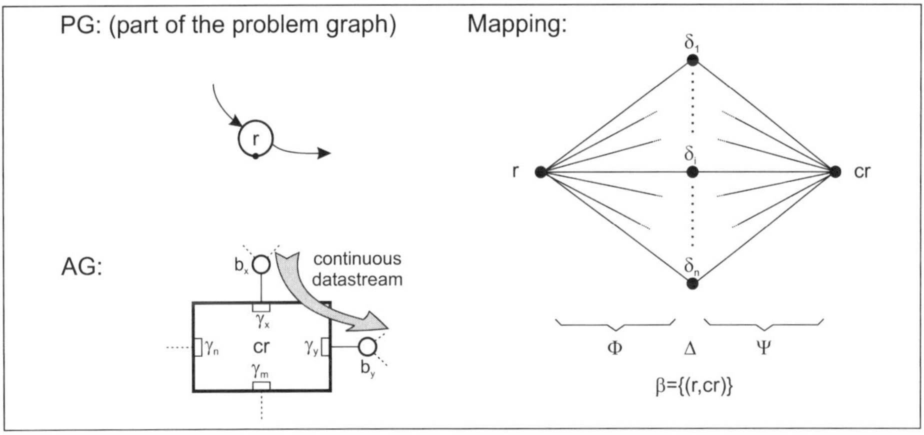

Figure 3.31: Continuous data stream on run-time reconfigured devices.

Now, assume that cr has to implement a set of configurations $\Delta = \{\delta_1, ..., \delta_n\}$. To ensure the routing from interface γ_x to γ_y during each configuration, task r has to be associated with every configuration, i.e., r is shared among the configurations. Subsequently, each implementation of a configuration has a copy of task r. Note that the continuous data stream is being interrupted during a configuration switch.

Example 6 (Field programmable port extender) *IP (Internet Protocol) packet processing systems are increasingly demanding more computing performance to keep pace with rising bandwidth requirements. Usually, hardware components (often ASICs) implement the performance critical tasks. The use of reconfigurable computing resources ensures the flexibility to implement new features and still provide a high computing power. In Lockwood [LNTT01] a field programmable port extender (FPX) has been presented enabling reconfigurable packet processing functions between a network* switch *and a corresponding* line card *to upgrade the system with new functionality (see Figure 3.32). However, no formal model has been given for modeling the run-time reconfiguration of the system which is now subject of this example.*

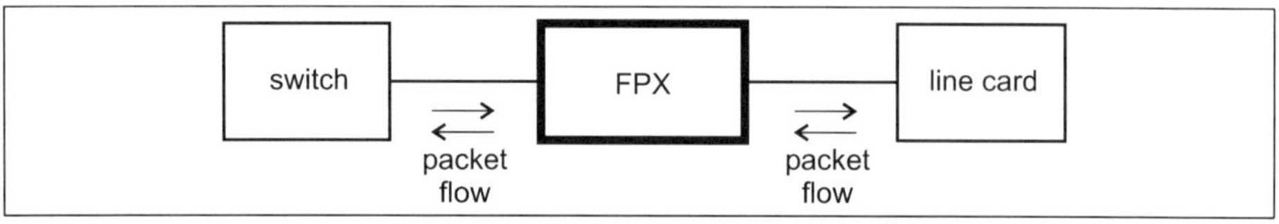

Figure 3.32: Embedding of the FPX.

A model for the FPX (see Figure 3.33) is considered using the proposed RPS formalism. The architecture graph consists of two connected FPGAs NID *(network interface device) and* RAD *(reprogrammable application device). The NID is responsible for transmitting and receiving packets from/to the switch/line card. Furthermore, it is the host computing resource for the RAD and enables run-time reconfiguration of the RAD resource. The RAD supports partial reconfiguration for the implementation of two concurrent configurations. Therefore, it is modeled hierarchically consisting of two basic FPGA architecture components* $\{Mod_A, Mod_B\} \in C^B$. *Each of these resources has access to non-shared off-chip memory, e.g.,* Mod_A *has access to* $SDRAM_A$ *and* $SRAM_A$. *The memory PP models the RAD's programming port to load a new configuration.*

The problem graph covers the functionality as mentioned in [LNTT01]. The most important part of the NID is the four port switch s *that reads data packets from four input queues* $q_1, ..., q_4$ *and forwards them to output queues* $q_5, ..., q_8$ *according to the information provided by* vc. vc *is a* virtual circuit lookup table *that relates data packets and switch ports. The* RAD *is able to host two concurrent configurations, e.g.,* δ_1 *and* δ_4. *Each of them receives/sends data from/to the task* s *via one intermediate*

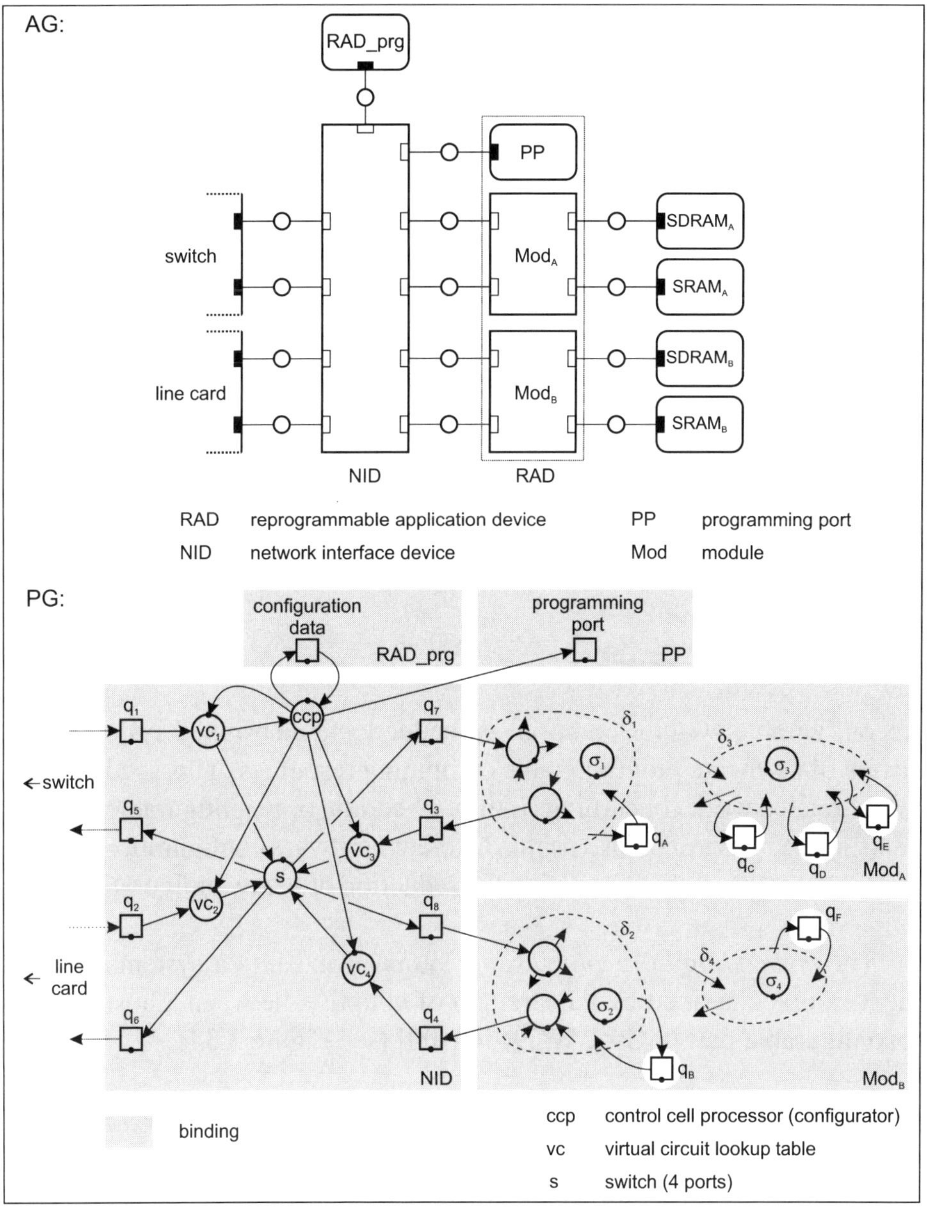

Figure 3.33: Field programmable port extender (FPX) [LNTT01].

queue. Furthermore, each configuration may have buffers, e.g., q_A, that are bound to the connected memories $SDRAM_1$, $SDRAM_2$, $SRAM_1$, or $SRAM_2$ respectively.

Task cpp (control cell processor) is addressed via special control cells sent by the switch. It is responsible for the content of the virtual circuit lookup tables and supervises the reconfiguration of the RAD. RAD configuration data sent to cpp are stored in the memory RAD_prg. After transmitting a whole configuration, cpp forwards the data to the programming port of the RAD which reconfigures the FPGA.

As visible in the model, there exists no connection between cpp (configurator) and any of the configuration's dispatchers. Therefore, the present implementation described in [LNTT01] does not support reply messages between the dispatcher and the configurator which prevents dynamic scheduling (see Section 3.3.1). However, a connection between configurator and dispatcher is desirable. For this purpose, we suggest to insert two additional buses; one between NID and Mod_A *and another between NID and* Mod_B *(not shown here). Using these two buses the considered communication could be easily established.*

3.3.9 Related Work

Reconfigurable systems comprise embedded and distributed systems consisting of hardware programmable computing resources. This section elaborates related work concerning (i) the embedding of reconfigurable components in their environment, (ii) the different hardware configuration modes, and (iii) optimizations concerning the reduction of the reconfiguration time.

The *embedding of reconfigurable system parts* into a system environment can be classified by the strength of coupling between a host and the reconfigurable part [GG95, WC96, CH00] (see Figure 3.34).

The most intricate form is a *reconfigurable resource* (see *rr* in Figure 3.34) built together with a general/special purpose processor core to speed-up the most performance critical instructions of an application, e.g., [WC96, HFHK97]. It provides a high speed-up due to the tight coupling via common processor registers.

The *coprocessor* approach attaches reconfigurable logic to a local processor bus or dedicated I/O ports of a host processor by using FPGA devices [RLG+98, HBS98, ETT98] or reconfigurable on-chip area [BG00, XILb, Tri, Kea00, Inc]. The coprocessor is configured by the host, and

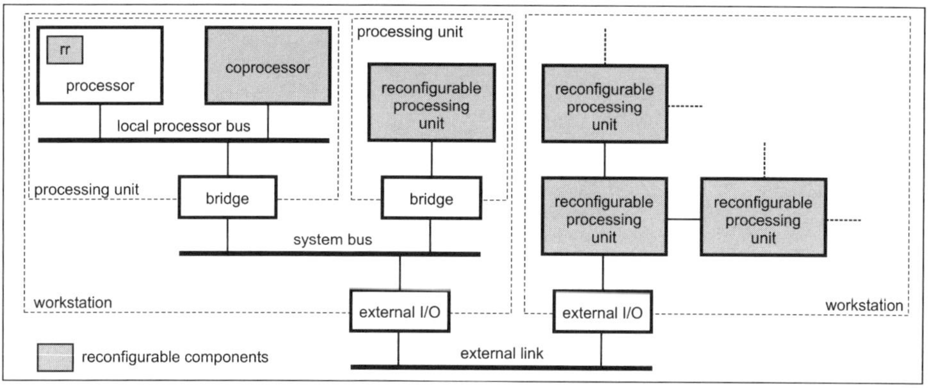

Figure 3.34: Coupling of reconfigurable system parts.

performs larger parts of the application, e.g., in form of function calls. Usually, the host and the coprocessor operate in parallel and exchange only intermediate results. Careful problem graph node partitioning between host and coprocessor reduces bandwidth requirements [TBT97]. However, interface circuitry dealing with I/O pin assignment and protocol translation is required [HB97, ET98a].

An even weaker form of coupling provides reconfigurable logic as *reconfigurable processing unit* in a multi-processor environment. Add-on boards host one or several FPGAs and are linked via the system bus to the reminder of the system. Corresponding approaches include SPYDER [IS95], Pamette [Sha], COSYMA [EHB$^+$96], SUNDANCE [Sun], etc. These systems are used, e.g., for rapid prototyping [HBKG98], acceleration of processing intensive tasks [LSS99], and generally provide an API (Application Programming Interface) on the host for ease-of-use [BH98, MMF98]. The communication overhead is reduced by executing large parts of the application at task granularity.

A *stand-alone reconfigurable system* (e.g., a workstation) is the weakest form of coupling where the reconfigurable part is a multi-FPGA system and performs most of the (or even the complete) application without any host communication. These systems can be viewed as hardware supercomputers and are successfully used, e.g., as ASIC emulation systems [Apt, Qui].

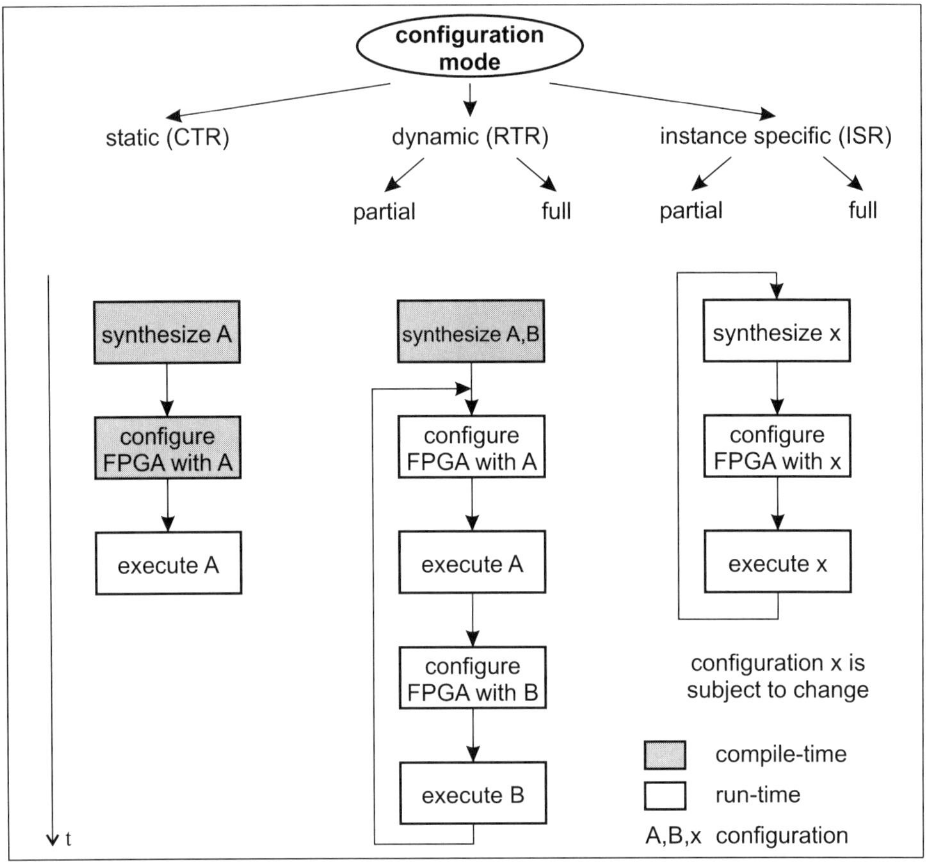

Figure 3.35: FPGA configuration modes.

For the *configuration of a reconfigurable computing resource* three different modes are known (see Figure 3.35):

Static (or **C***ompile-***T***ime)* **R***econfiguration* (**CTR**)
CTR resources have exactly one configuration during the lifetime of the application [HW95]. During compile time, a resource's configuration is synthesized and used to configure the FPGA. During run-time, the configuration is executed. The goal is to combine ASIC performance with processor flexibility [BP95]. Such resources are used for logic emulation [PB98, Qui] and rapid prototyping systems [KG97]. Here, FPGAs quickly implement hardware functions for validation and design space exploration.

Dynamic (or **R***un-***T***ime)* **R***econfiguration* (**RTR**)

Dynamic (or run-time) reconfigured resources have a set of time exclusive configurations. These configurations are synthesized during compile-time. During application execution the resource's configuration is repeatedly switched [HW95, SSH$^+$99, KV98, PB99]. The goal of dynamic reconfiguration is to optimize and reuse the available hardware resources concerning unused circuitry and optimally adapted implementations [LM95, SJV95, GL99]. Such resources provide the allocation of a fast hardware implementation depending on run-time conditions [SB94, WH96, EP00a], e.g., for packet processing of network processors [LNTT01]. Further applications include embedded systems where frequent hardware updates are expected, e.g., network switches that can be reconfigured to support different topologies [AH95].

I*nstance-***S***pecific* **R***econfiguration* (**ISR**)

The configuration data of instance-specific reconfigured resources are unknown at compile-time. Problems just arise during run-time. Therefore, synthesis, resource configuration, and execution are part of the overall application. The goal of instance-specific reconfiguration is to accelerate computation intensive parts that arise at run-time. This strategy has been successfully used to accelerate decision problems [Pla00].

As the configuration switch of a reconfigurable resource usually involves a considerable amount of configuration data (about 766kB of data for a XILINX XCV1000) reconfiguration time is a performance limiting factor for RTR/ISR resources. Therefore, various *optimization techniques* have emerged to reduce the necessary configuration data. Subsequently, RTR/ISR systems further divide into fully and partially reconfigurable systems [Mar99] (see Figure 3.35). Full reconfiguration involves reprogramming of the complete resource, whereas partially reconfigurable resources provide the alternation of selective parts only. Further research issues for full reconfiguration include multi-context FPGAs [WH96, TCJW97, SV98] that store several configurations on the same chip, configuration data compression [HLS98] making use of hardware supported decompression (XILINX XC6200), and optimal partitioning of communicating nodes into temporal/spatial exclusive configurations to reduce the number configurations [KV98, CV99, PB99, LCD$^+$00]. The proposed approaches may be applied to partially reconfigurable devices as well. However, it has not been focus of research yet.

3.4 Summary

This chapter provides a hierarchy of specification models to capture and describe synthesis problems for dataflow oriented heterogeneous systems. Essentially, the focus of the models is on the specification of communication channels between (i) problem graph nodes, and (ii) on communicating target components.

The base model GPS (**G**eneral **P**roblem **S**pecification) is parent of the proposed formalisms and provides definition of terms, basic elements, and composition rules to specify a dataflow problem for synthesis. It comprises a problem graph describing the target independent system behavior, an architecture graph capturing the target platform, and a mapping between problem and architecture graph.

EPS (**E**mbedded system **P**roblem **S**pecification) is an extension of GPS and targets at the specification of synthesis problems of dataflow oriented embedded systems consisting of general and special purpose processors, programmable as well as dedicated hardware components, and memories.

The RPS (**P**roblem **S**pecification for **R**econfigurable systems) model includes the specification features of EPS and additionally supports runtime reconfigurable hardware components.

The following features of the proposed models are new and not part of known models:

- The formalisms base on each other and provide rising modeling capability. Formalisms, nodes of the problem graph, and components of the target architecture can be represented in UML notation. This property assures an efficient object-oriented implementation of a synthesis tool.

- The specification of the communication is an integral part of the suggested formalisms. The behavior model denotes *necessary* communication relations between interacting nodes. The target architecture provides *available* communication resources and the mapping specifies which resource will implement which communication relation.

- The components of the target architecture explicitly provide two different interface types considering the connection features of hardware and software programmable components.

- The novel RPS model provides a set of application scenarios. Problem graph nodes can be migrated between configurations associated with the same or adjacent computing resources. Virtual configurations provide a way to replace complete configurations on a "pin-compatible base".

Chapter 4

Model Refinements

Based on the models introduced in the last Chapter 3, appropriate refinements and implementations of EPS and RPS models are presented (summary shown in Figure 4.1). At first, refinements of point-to-point communication channels are elaborated (Section 4.1). Thereby, additional nodes are inserted into a problem graph PG modeling the access to communication channels. Subsequently, these nodes lead to a refined problem graph rPG as well as to a refined mapping. The arising communication types are summarized in a taxonomy of communication types. Furthermore, an object-oriented model is introduced to specify the architecture graph (Section 4.2). Based on this model, it is shown how automatic synthesis of interfaces and device drivers works. Next, node implementation issues for problem graph nodes are elaborated (Section 4.3). Furthermore, implementation approaches concerning the dynamic FPGA reconfiguration are outlined (Section 4.4). A short summary is given in Section 4.5.

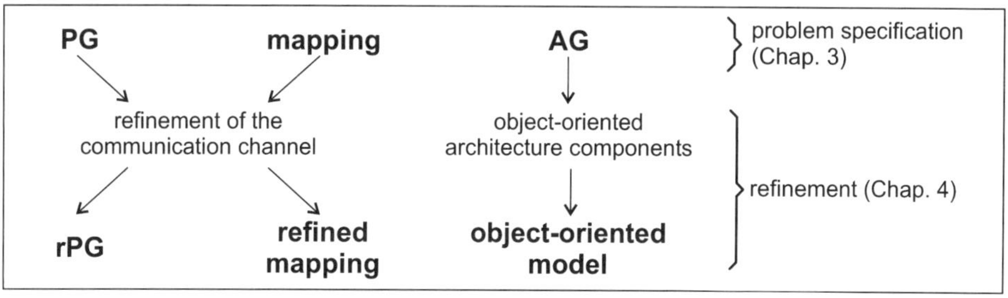

Figure 4.1: Chapter outlook.

4.1 Communication Channel

The proposed formalisms base on directed point-to-point communication channels between sender and receiver nodes. A channel is the implementation of a directed edge of the problem graph and consists of connected architecture components, appropriate device drivers and interface circuitry enabling the data transfer. It is assumed that a channel provides *no memory*. Instead, memory is modeled explicitly by buffer nodes in the problem graph. This section elaborates the *refinement* of a problem graph edge which includes the channel *access semantics* as well as a *taxonomy* of communication types.

4.1.1 Refinement

To establish a communication channel between a sender and a receiver the corresponding edge in the problem graph has to be refined. Quite obviously, the goal is to find an appropriate implementation for each edge of the problem graph. Our proposed refinement consists of the insertion of *communication nodes* into the problem graph. These additional nodes represent parts of the target architecture that implement the final communication channel. During synthesis, these nodes are replaced by dedicated send/receive primitives that are executed for data communication. Generally, one or two communication nodes are inserted:

Single communication node
Each problem graph edge is replaced by two edges and a single communication node. Communication nodes are bound to intermediate buses between communicating architecture components. Such refinements have been used by Teich et al. [TBT97] in order to analyze the influence of communication time on the system performance. Figure 4.2a) shows a refined problem graph PG^* for the problem discussed in Figure 3.6a). A single communication node has been introduced between the communicating tasks. The dotted lines denote the node's binding to architecture components. For example, the communication node c_1 is bound to the bus b_1. However, using just a single communication node is not very well suited for the specification of an implementation as several architecture component interfaces can be involved. Subsequently, a more fine-grain refinement is aspired.

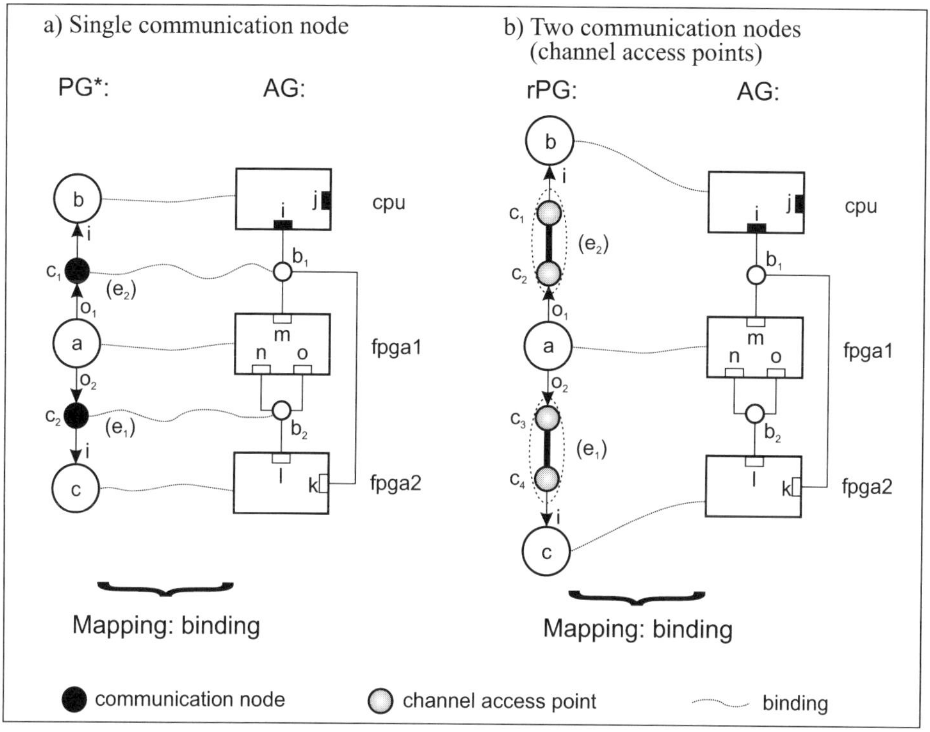

Figure 4.2: Refinements of the problem graph discussed in Figure 3.6a).

Two communication nodes

The proposed formalisms provide a more sophisticated refinement that is better suited for implementation [ETT98]. For each edge of the problem graph PG two communication nodes may be inserted. These additional nodes are implicitly associated with the involved components' interfaces by the specification of the edge binding τ. In the proposed terminology, such nodes are called *channel access points*. As an example, Figure 4.2b) shows a refined problem graph rPG of the problem discussed in Figure 3.6a). Here, two channel access points have been inserted per edge. The nodes model the communication between the three computing resources cpu, *fpga1*, and *fpga2*. During the synthesis process, channel access points are replaced by device drivers and interface circuits. For example, the binding of edge e_1 (edge between nodes a and c in the problem graph) $\tau(e_1) = (e_{n-b_2}, e_{l-b_2})$ in the specification implies that the channel access point c_3 is associated with interface n, and c_4 is associated with interface l. Bold unidirectional edges visually denote the communication and interaction between channel access points. Thin directed edges visually denote the communication between executable/buffer nodes and channel access points. Figure 4.3 shows the embedding of channel access

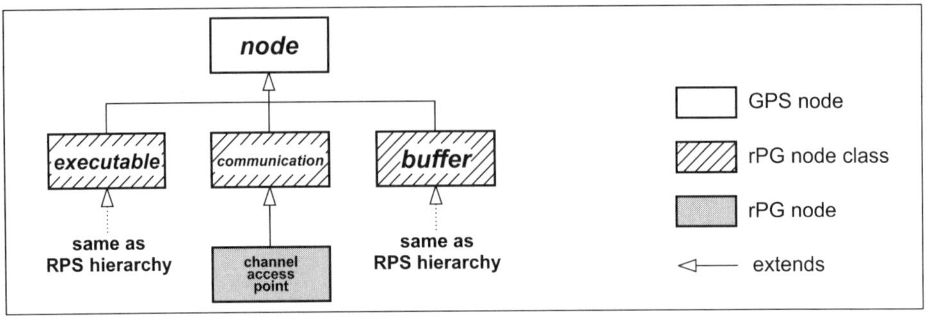

Figure 4.3: Node hierarchy of the refined problem graph (UML).

points into the node hierarchy. An rPG has the additional node class *communication*. *Channel access point* is an rPG node modeling the access to a channel.

The following elaborations outline the four relevant cases concerning the refinement of a single edge (see Figure 4.4):

a) *Two adjacent computing resources*
This case arises if two computing resources are adjacent and each of them has a bound node. Here, two channel access points are necessary (see Figure 4.4a). The resulting implementation of the edge is called *off-chip communication* as the data have to be transmitted via an external bus.

b) *A single processor computing resource*
Here, the two communicating nodes are bound to the same sequential processor computing resource (see Figure 4.4b). For the intermediate edge two channel access points are necessary, again providing send/receive of data items. The implementation is called *on-chip communication* as the data transfer takes place on the same computing resource.

c) *A single FPGA computing resource with incompatible node port protocols*
This case arises, if two communicating nodes are bound to an FPGA computing resource and their ports have incompatible protocols (as shown in

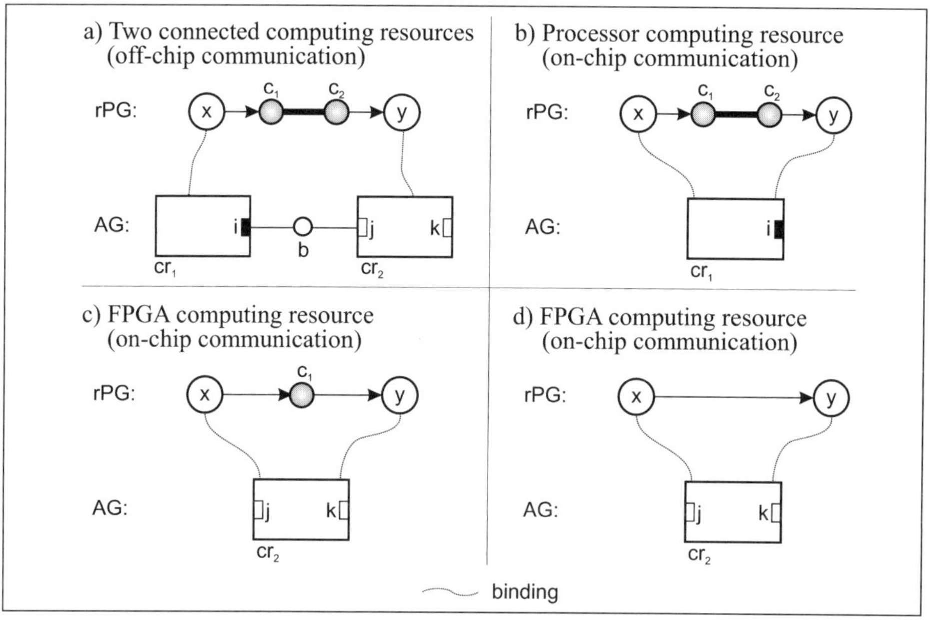

Figure 4.4: Valid refinements of an edge.

Figure 4.4c). In this situation, one intermediate communication node is necessary modeling the on-chip protocol translation (transducer circuit) between these two nodes.

d) *Single FPGA computing resource with compatible node port protocols*
This is the simplest case. Both communicating nodes have compatible protocols. Here, no communication node is required to implement the on-chip communication (see Figure 4.4d).

However, if (i) one of the communicating nodes is a buffer and (ii) one of the connected architecture components is a memory, a special refinement is applied. This issue arises due to different implementation techniques of tasks and buffers and is elaborated in Section 4.3.3.

The above considerations about channel access points can be seen from another point of view as well. In literature, the term *wrapper* denotes a circuit or a piece of software converting the interface of a module, function, block, etc., into another interface [GHJV95]. In our context, channel access points are wrappers for the problem graph nodes to access communication channels [ETT98] (see Figure 4.5a). The implementation of channel access points and the connected architecture components is denoted as *communication infrastructure*. The shaded area in Figure 4.5b) denotes the communication infrastructure of Fig 4.2b). Bold lines outline the implementation of the problem graph edges.

Definition 25 (Communication Infrastructure) *The communication infrastructure of an embedded system abstracts from the underlying channel implementation and comprises necessary hardware and software drivers to establish the communication channels between the problem graph nodes. Channel access points provide the sending/receiving of data items between the connected nodes.*

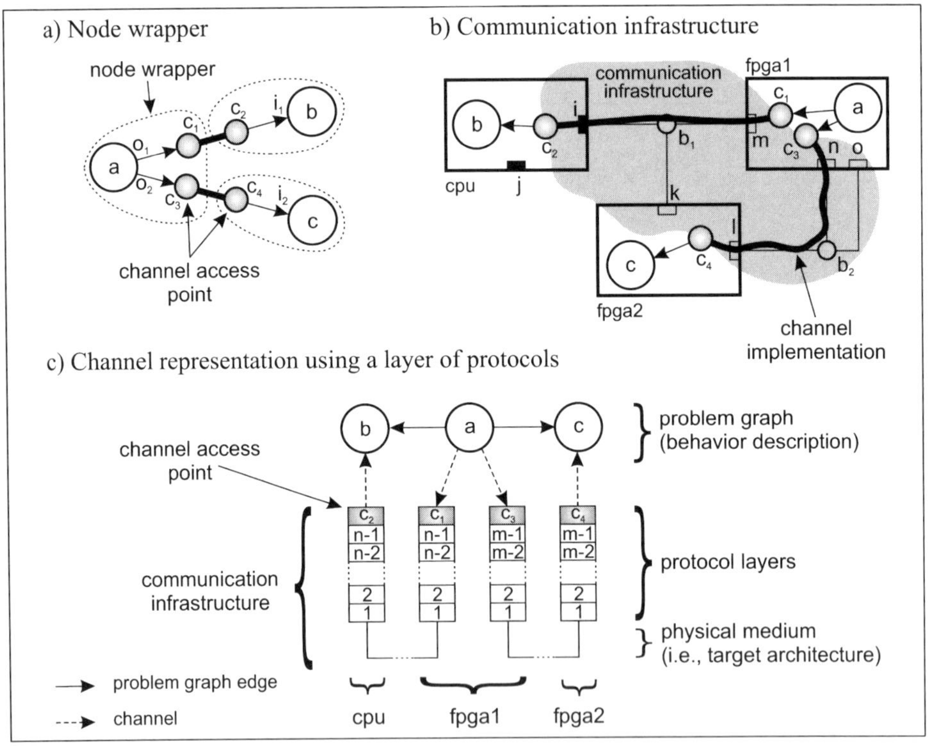

Figure 4.5: Channel refinement.

Our model of a channel access point assumes that the communication infrastructure can be represented as a series of adjacent protocol layers (protocol stack) [Tan89]. Each layer p offers services to the next higher level $p + 1$ based on the services of layer $p - 1$. The topmost layer is represented by the already discussed channel access points (see Figure 4.5c). The lowest layer is the target architecture of connected components. Intermediate layers (if any) may incorporate a computing resource's operating system that provides appropriate communication primitives. This concept of protocol layers has been introduced by the OSI reference model [DZ83] and is being used in a number of co-design environments, e.g., [OB98, EP00c].

4.1.2 Channel Access Semantics

The GPS formalism supports *blocking* and *non-blocking* channel access points which is sufficient to support synthesis problems for marked graphs, FunState components, etc. (see Section 3.1). For example, for marked graphs blocking communication is appropriate. In general, an access point provides one or more primitives that encapsulate an appropriate protocol stack. The actual data transmission consists of three phases:

1. Initialization I
The data transmission is being initiated by executing a communication primitive. A typical task comprises appropriate register settings of the involved computing resource's interface.

2. Data transmission T
The data item(s) are transmitted. This usually involves computing resources, dedicated controllers, device drivers, etc.

3. Completion C
Phase 3 ends data transmission and comprises (i) *test* for completion **CT**, and (ii) *wait* for completion **CW**.

$$\mathbf{CT} = \begin{cases} true & : \text{ transmission completed} \\ false & : \text{ else} \end{cases}$$

Note that **CW** is equivalent to **while (not CT)**. A *blocking* channel access is an atomic operation, i.e., the sequence $\mathbf{I} \rightarrow \mathbf{T} \rightarrow \mathbf{CW}$ is indivisible from a caller's point of view. A *non-blocking* channel access is a non-atomic operation. The communication is merely initiated by phase **I** which subsequently issues phase **T**. To finalize the communication **CT** or **CW** have to be issued separately.

Non-blocking communication provides a more general approach than blocking communication as a blocking access can be established using non-blocking primitives but not vice versa. An implementation using non-blocking communication is usually faster than an implementation using blocking communication due to the potential overlapping between node execution and data communication [Pac97]. Furthermore, during a transmission phase **T**, nodes may continue their processing but may undergo a degradation in available CPU processing power. This may arise due to an involvement of a node's computing resource for the requested data transmission [NTE99].

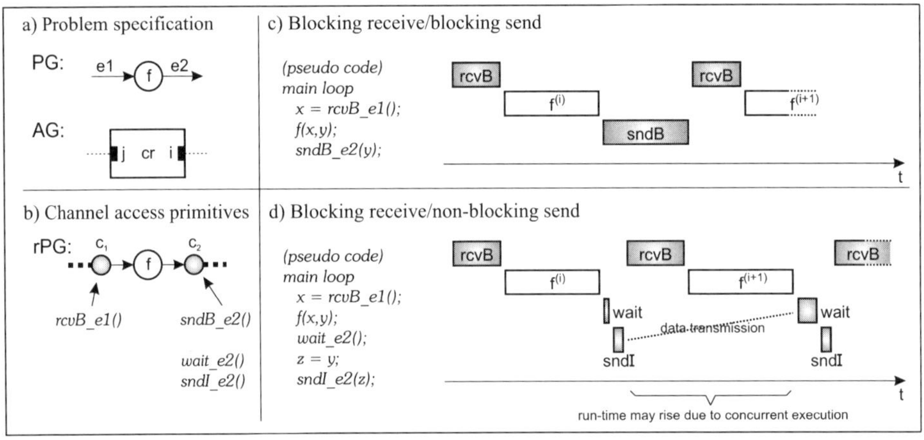

Figure 4.6: Comparing blocking/non-blocking channel access points.

As an example, consider the problem specification in Figure 4.6a) consisting of a task f bound to a processor computing resource cr. During the refinement, the synthesis process for the edges e_1 and e_2 provides channel access primitives as shown in Figure 4.6b). Figure 4.6c) outlines an implementation and execution of the specification if blocking access (*sndB, rcvB*) is used for both channels. The blocking access points enforce a sequential execution: receive data on edge e_1, execution of task f, and send data on edge e_2. In contrast, a non-blocking send (*sndI (send immediate)*) via channel e_2 enables concurrent execution: sending the output value of $f^{(i)}$ can be concurrent to receiving a new data for $f^{(i+1)}$ and executing $f^{(f+1)}$ (see Figure 4.6d). However, the computing resource cr has to support such a parallelism, e.g., by a built-in DMA module. Furthermore, the concurrent execution may extend the run-time of the individual parts as indicated in Figure 4.6d).

Note that in literature often *synchronous communication* is a synonym for blocking communication and non-blocking communication is a synonym for *asynchronous communication* [BA90].

4.1.3 Taxonomy of Communication Types

O'Nils et al. [OJ97] suggested nine distinct architectural communication routes between software processes, hardware processes, and library modules on the same or different chips (no buffers considered within their problem specification). In contrast, a taxonomy is proposed that bases solely on the binding of problem graph nodes to basic architecture components and the existence of one or several computing resource configurations [EPT99]. It emerged that the communication routes in [OJ97] represent just two cases in the taxonomy because software and hardware implementations are considered as equivalent for the synthesis of communication channels (see Section 4.2).

Figure 4.7a) shows the problem specification considered to establish the taxonomy. A problem graph consisting of two tasks x and y and a buffer b has to be implemented on a target architecture comprising two FPGA computing resources cr_1 and cr_2 and a memory mem. The first refinement step introducing two communication nodes on each edge ($c_1, ..., c_4$) is outlined in Figure 4.7b).

Depending on the binding β our taxonomy differentiates between four types of communication (enumeration in rising implementation complexity):

a) *On-chip communication*

On-chip communication arises if both nodes of an edge are bound to the same basic architecture component, e.g., $\{(source(e_1), cr_1), (sink(e_1), cr_1)\} \in \beta$. This is the simplest case, as this type of data exchange is supported by most of the computing resource's programming languages such as C or VHDL. Therefore, channel access primitives use the communication facilities provided by the programming languages or the operating system, e.g., a shared variable.

b) *Off-chip communication*
Off-chip communication arises if both nodes are bound to different basic architecture components, e.g., $\{(source(e_2), mem), (sink(e_2), cr_2)\} \in \beta$. Here, device drivers and/or interface circuits are necessary to access the corresponding interfaces of the involved computing resources. In the special case where one of the two architecture components is a memory, only one device driver (or interface circuit) is necessary to implement the edge (see Section 4.3.3 for more details). In the example above, access point c_3 is not required as a memory does not need a device driver. Furthermore, off-chip communication involves the assignment of device pins to a channel implementation.

c) *Interconfiguration communication [HW95, EP00c]*
Interconfiguration communication arises if exactly one node of an edge is bound to a computing resource that undergoes several reconfigurations. As an example, $\{(source(e_2), mem), (sink(e_2), cr_2)\} \in \beta$, $|\Delta^{cr_2}| > 1$, where Δ^{cr_2} denotes the set of configurations of cr_2. In this case, the nodes communicate over configuration borders. Subsequently, a channel is being physically disrupted during reconfiguration. This communication type requires additional control logic in the implementation to assure correct data transmission. As an example, Figure 4.8a) shows a problem graph PG and its refined problem graph rPG where cr_2 has two configuration δ_1 and δ_2. As a rule of thumb, if an edge is crossed by exactly one configuration border in the problem graph PG (visual notation) the resulting communication type is interconfiguration communication. Note that his communication type may also arise if one of the computing resources is a processor resource.

d) *Interconfiguration communication with open channel* [EP00c]
Interconfiguration communication with an open channel arises if both nodes of an edge are bound to adjacent basic computing resources that undergo several reconfigurations, e.g., $\{(source(e_1), cr_1), (sink(e_1), cr_2)\} \in \beta$, $|\Delta^{cr_1}| > 1$, $|\Delta^{cr_2}| > 1$. Here, an affected channel may be physically disrupted by both, sender and receiver node. Figure 4.8b) outlines a corresponding example. As a rule of thumb, if an edge is crossed by two configuration borders in the problem graph PG (visual notation) the resulting communication type is interconfiguration communication with open channel.

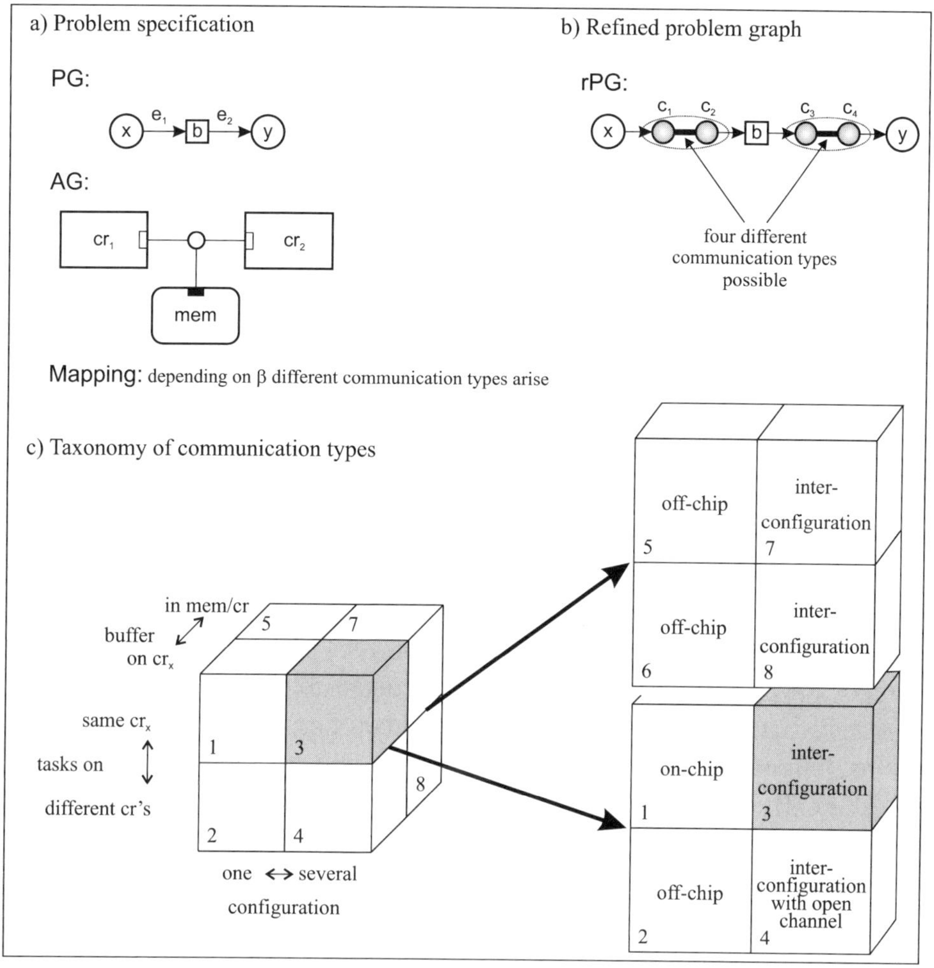

Figure 4.7: Taxonomy of communication types.

Figure 4.7c) shows the proposed taxonomy as cube representation outlining the eight conceivable binding cases. The three orthogonal axes denote (i) if the tasks x and y are bound to the same or different computing resources, (ii) if the buffer is bound to one of the computing resources or the memory, and (iii) if the involved computing resources undergo one or several configurations. The cases 1 to 8 in the cube show the most difficult communication type for a certain implementation. As an example, case two is considered. If $\beta = \{(x, cr_2), (b, cr_2), (y, cr_1)\}$, $|\Delta^{cr_1}| = 1$, $|\Delta^{cr_2}| = 1$, then *on-chip communication* arises for the implementation of edge e_1 and *off-chip communication* arises for the implementation of edge e_2.

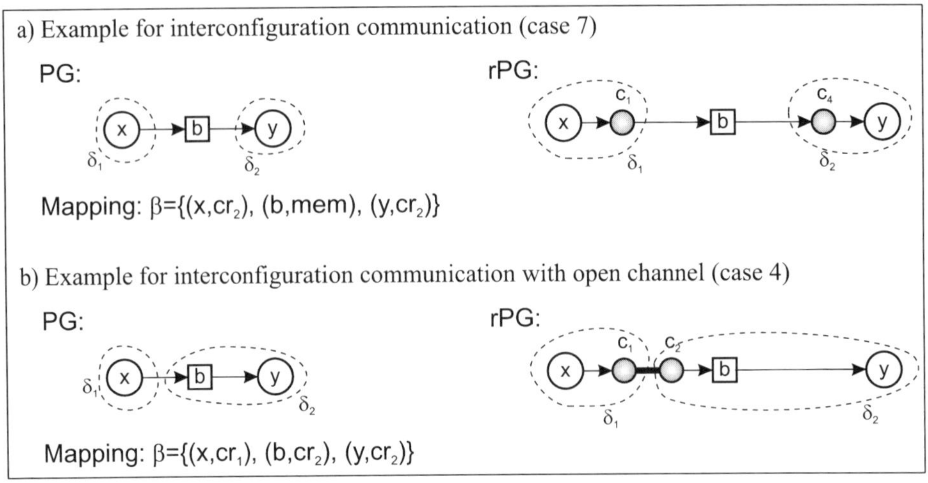

Figure 4.8: Interconfiguration communication.

In general, embedded systems without reconfigurable computing resources have only *on-chip* and *off-chip communication* (cases one, two, five, and six in Figure 4.7c). The same holds for embedded systems where the reconfigurable computing resources have only one configuration (CTR systems). By comparison, the communication routes proposed in [OJ97] are just represented by the cases one and two. If computing resources have several configurations (like in RTR systems) nodes can communicate over configuration borders. Normally, intermediate buffers are assigned to memories (cases seven and eight) to hold data between subsequent configurations which requires *interconfiguration communication.* However, if the problem graph does not include such buffers or the buffers are assigned to computing resources that undergo reconfiguration, either *interconfiguration communication with open channel* (case four) appears or a partially reconfigurable computing resource is required were dedicated non-reconfigured area is necessary to implement the buffer (case three).

As an example, the refinement of virtual configurations is elaborated that have been specified in Figure 3.30. As shown in Figure 4.9 channel access points have been inserted to model the interconfiguration communication between subsequent configurations. Related pairs of channel access points are denoted with c_x and c'_x, respectively, e.g., c_1 and c'_1. Edges that are connected to the same port have the same interface circuitry in each configuration. Subsequently, the channel access points have the same name. For example, the two edges connected to the output port o of configurator ρ have the same channel access points c_2 and c_2', respectively.

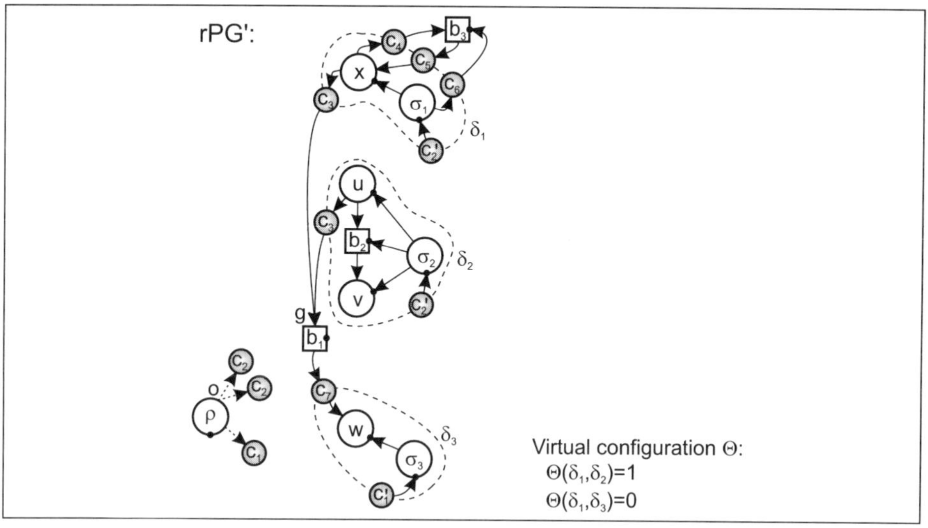

Figure 4.9: Refined problem graph of Figure 3.30.

4.2 Architecture graph

Prospective interface circuitry and device drivers are modeled by communication nodes in a refined problem graph. As discussed in Section 2.2.2, there exist a lot of methodologies to actually create these code fragments. It is outlined that some approaches can require a considerable amount of manual work and may be restricted to certain platforms. In this thesis an object-oriented approach is proposed that is able to support and combine several methodologies as well as 3rd party tools to generate an appropriate communication infrastructure. This section introduces the object-oriented modeling of architecture graph components. Based on these components the process of automatic communication synthesis is elaborated.

4.2.1 Object-oriented Component Model

The principal focus of the proposed object-oriented approach is the architecture graph that models a set of connected devices. In a running system implementation, these devices communicate among each other to implement the specified communication channels. Essentially, for each component in the architecture graph there exists an object-oriented counterpart, called *chip model* [EP00c]. A chip model captures important properties of the actual device like the number and type of I/O modules. However, the proposed approach of modeling is not limited to communication only but may as well include the modeling of peripheral parts such as timers. Furthermore, chip models include estimation data about the prospective implementation such as the expected communication speed per channel and implementation overhead.

The proposed object-oriented approach is outlined by elaborating the modeling of the hierarchical computing resource Triscend A7 [Tri] (see Figure 4.10). It is assumed that an *integrated circuit* has a *block diagram* (taken from a corresponding device datasheet) that denotes the most important internal blocks (see Figure 2.4 for an enlarged version of the block diagram). For example, the most important blocks of the A7 device are: (i) the *ARM7* processor core providing a sequential computing resource (*processor*), and (ii) the configurable system logic *CSL*, a hardware programmable computing resource (*FPGA*). These two blocks are summarized in a corresponding chip model of the *A7* device (bold box). In fact, chip models are classes that represent the actual device. They are composed of a set of further classes by using object-oriented modeling techniques such as polymorphism, class inheritance and object composition. The static relations between these classes are visualized in a *class diagram*. Cursive face denotes *abstract classes* describing common properties. For example, each *communication block* class provides an estimation about its transfer rate. Regular face denotes *concrete classes* that can be instantiated. The class diagram is divided into three sections:

Component Models
Component models are abstractions of actual devices and describe common component properties such as the number of configurable logic blocks of an FPGA computing resource.

Chip Models
Chip models exists for components that may appear in an architecture graph. A chip model describes the main features of an actual device. For example, the *A7* chip model is a hierarchical component. It is composed

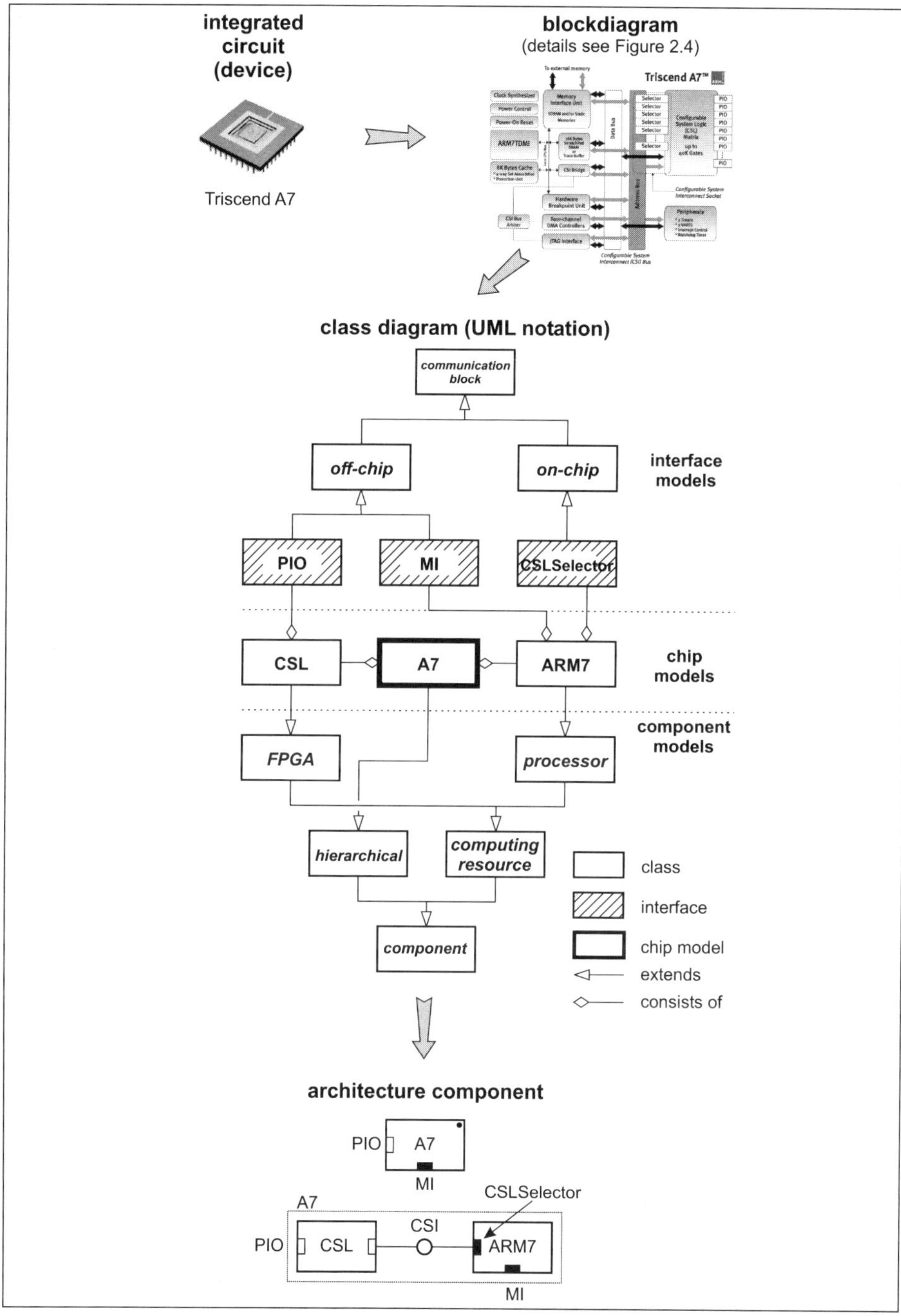

Figure 4.10: Modeling the Triscend A7 computing resource.

of *CSL* an FPGA computing resource as well as *ARM7* a processor computing resource. The structural representation is captured by an internal architecture graph.

Interface Models
Interface models describe the interfaces of actual devices. The top classes describe common properties, e.g., *off-chip* interfaces have to deal with device pins. Each interface class (hatched boxes) comprises a set of configurable interface generators. These generators actually generate device drivers and interface circuitry with different protocols for the component's interfaces. Generators include the discussed approaches such as library-, template-, component-, generator-, and platform based. They can deal with standard as well as proprietary protocols and interfaces. As an example, the *ARM7* has two interfaces: (i) *MI* enabling to communicate externally via the memory interface, and (ii) *CSLSelector* providing the on-chip data transfer between the *CSL* and the *ARM7*. The *CSL* has a parallel I/O interface *PIO*.

Subsequently, a chip model of a device contains all necessary information to generate interface circuitry for its R-type interfaces and device drivers for its P-type interfaces, respectively. It provides a very flexible approach to cope with the targeted heterogeneous architectures. Comparing with related approaches [VT97, BHLM91, COB95a] the following main *advantages* can be seen:

Simple retargeting of the interface generation
The use of *unified class interfaces* provides the exchangeability of components in the architecture graph (see next Section 4.2.1). It enables the suggested approach to be used as a backend in an overall embedded system designflow. For example, changing the type of a processor computing resource to an FPGA computing resource requires only the change of the component type in the architecture graph. Furthermore, changing the connection structure of the components entails just the appropriate structural modeling by the architecture graph.

Reuse
The interface classes represent a set of configurable interface generators that can be applied for different chips. This property is especially useful to

model derivatives of microcontroller and DSP families. Essentially, such devices quite often use a set of common blocks but in different compositions. However, the use of common blocks is the foundation of platform based design that is supported through the suggested reuse property.

Simple device modeling
The unified class interfaces provide a simple composition of chip models. Furthermore, they enable the modeling of components that do not exist as actual devices yet but would provide the best choice for a problem under consideration. This can be a starting point for a special purpose device implementation. The class interfaces include access to a database of estimation data as well. In fact, there exists a *component selection guide* outlining the most important chip properties such as the supported interfaces, performance characteristics, etc.

Object interaction
Chip models interact and exchange messages. This property considerably simplifies the synthesis process of the communication infrastructure (see Section 4.2.3). For example, remember that FPGA computing resources have R-type interfaces and require appropriate interface circuitry to be connected to other components. For each component that can be connected to an FPGA component one could write a corresponding interface generator and add it to an FPGA chip model. Subsequently, such an FPGA chip model would grow bigger and bigger and had to be adapted to every new component. Furthermore, there exist different FPGAs distinguishing in speed, available programmable area, vendor, etc., for which chip models had to be updated. Fortunately, by using the *object interaction* property chip models are much simpler. Namely, an interface generator of a connected processor or memory component has the ability to generate several code fragments. One such code fragment can represent an interface circuit that is transferred to a connected FPGA component. This circuit provides the appropriate connection between the FPGA component and the processor/memory component (see Section 4.2.3).

The outlined advantages above are paid by the following main *disadvantages*:

Object-oriented modeling required
For each device, a corresponding chip model has to be established and embedded into the compound of existing models and classes. Essentially, each model has to be compliant with the provided class interfaces.

Writing interface generators
Interface generators provide a very general concept to generate the communication infrastructure. However, depending on the type of interface they may include a rather large amount of source code. For example, an interface generator based on a library has to select and possibly configure predefined device drivers and interface circuits only. On the other hand, a template based generator may be composed of several sub-generators each providing a part of the final solution.

Unified Class Interfaces

Unified class interfaces include (i) the use of inherited properties, and (ii) predefined method signatures that describe object interaction. This section briefly outlines part of the unified class interfaces by the use of JAVA [AGH00] code examples. Again, part of the Triscend A7 computing resource is modeled (see Figure 4.10).

Prg. 1 shows the most important parts of the class *computing resource*. As shown on line *1* a computing resource is an abstract class that inherits the properties of a *component*. Line *2* denotes a dynamic array of references to available interfaces of a computing resource. The methods *create()* (lines *5 ... 7*) and *generate()* (lines *10 ... 12*) establish the component's internal data structures required for synthesis and enable the automatic interface synthesis process for this resource.

Prg. 1: Computing resource class (language: JAVA)

```
1  public abstract class ComputingResource extends
    Component {
2    protected Vector availableInterfaces = new Vector();
3
4    // create data structure for synthesis
5    final public boolean create( ... ) {
6      :
7    }
8    // start synthesis process of interfaces
9    final public void generate( ... ) {
```

```
        :
    }
}
```

The next code example Prg. 2 outlines that a *hierarchical* component has an internal architecture graph (line *2*). The architecture graph is being defined by a derived chip model.

Prg. 2: Hierarchical computing resource class (language: JAVA)

```
1  public abstract class Hierarchical
   extends Component {
2    protected Graph internalArchitectureGraph = new
     Graph();
3      :
4  }
```

Prg. 3 shows the *ARM7* class which is derived from a *Processor*. Note that ARM7 is a concrete class that can be instantiated. Lines *3* ... *12* denote the initialization phase of the object (i.e., *constructor* in terms of JAVA). Line *4* executes the constructors of the superclasses, i.e., constructors of *Processor*, *ComputingResource*, and *Component* in descending order. In line *5* processor properties such as available CPU processing power, available input/output device pins, or estimation about power consumption are read from a database. Lines *7* and *10* denote the instantiation of interfaces, i.e., a *CSLSelector* and an *MI* object. In lines *8* and *11* these objects are added to the set of available interfaces.

Prg. 3: ARM7 class (language: JAVA)

```
1  public class ARM7 extends Processor {
2
3    public ARM7(String InstanceName) {
4      super(InstanceName);
5      getPropertiesFromDatabaseProcessor(this);
6
7      CSLSelector cslSelector = new CSLSelector(this);
8      availableInterfaces.addElement(cslSelector);
9
10     MI memoryInterface = new MI(this);
11     availableInterfaces.addElement(memoryInterface);
12   }
13 }
```

The code excerpt Prg. 4 outlines the concrete class of the hierarchical component *A7*. Lines *3* ... *6* denote the initialization phase of the object. The structure of the internal architecture graph is read from a file. On that occasion, the necessary objects are instantiated and inserted into the architecture graph. This approach provides a very flexible object construction as the architecture graph can be changed just by rewriting a text file.

Prg. 4: A7 chip model (language: JAVA)

```
1  public class A7 extends Hierarchical {
2
3    public A7(String InstanceName) {
4      super(InstanceName);
5      internalArchitectureGraph = getAGFromFile(this);
6    }
7  }
```

A corresponding specification file is shown by Prg. 5, written in the script language CCSL [Eis99] (Communication Channel Synthesis Language). Lines *2* and *3* create the two internal computing resources *ARM7* and *CSL*. Line *6* creates the bus node *CSI*. In the lines *9* and *10* the two resources are connected to the bus. The *ARM7* is connected via its interface *CSLSelector*.

Prg. 5: A7 internal architecture graph specification (language: CCSL)

```
1  % computing resources
2  cr cpu  ARM7
3  cr fpga CSL
4
5  % bus
6  bus CSI
7
8  % connect resources to the bus
9  connect ARM7 CSI CSLSelector
10 connect CSL  CSI
```

As a brief overview, Figure 4.11 shows the coherence between chip models, CCSL scripts, and the proposed CCS framework (see Section 5.3). The *device database* contains the components that may appear in an architecture graph. The *IP database* contains predefined code such as source code for nodes of the problem graph. A *CCSL script* specifies the synthesis problem which includes the problem graph, the architecture graph, as well as a mapping. Furthermore, it contains commands for the control of the synthesis process of the CCS framework. The output of the framework are device drivers as well as interface circuitry that establish the requested communication infrastructure.

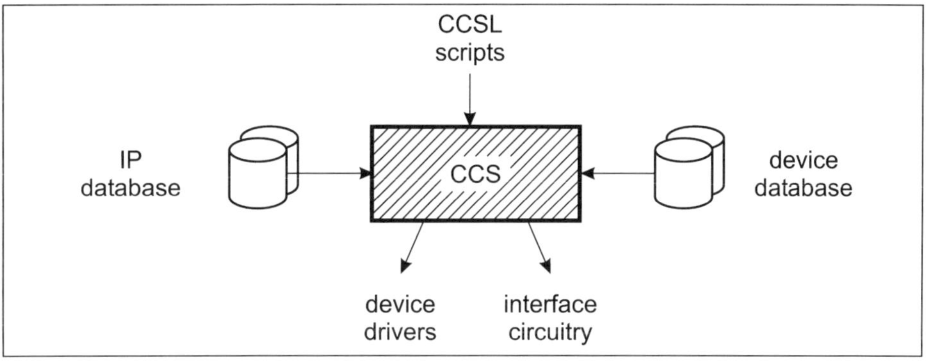

Figure 4.11: Brief overview: Framework CCS.

4.2.2 Component Wiring

To establish the communication between components, their interfaces are connected to common buses. Essentially, buses represent wires providing the physical data transmission. The wires themselves are connected to the device's pins. To enable the automatic synthesis of device drivers, interface circuitry, and their proper interconnection the actual physical wiring has to be considered. This is especially important for R-type interfaces as the generated interface circuits have to be connected to device pins.

To elaborate these issues the architecture given in Figure 4.12a) is considered. It consists of a processor computing resource cr_1, an FPGA computing resource cr_2, and a memory mem connected to a common bus b. Figure 4.12b) denotes the block diagram of the actual target wiring. Each device has a set of pins, e.g., cr_2 has pins $P4$, $P2$, $P7$, $P0$, $P3$, $P1$, and $P9$. Common wires connect the devices. For example, there is a common wire w_2 connecting device pins $cr_1.P1$, $cr_2.P7$, and $mem.P3$. All available wires of a bus and their connections to component pins are described in a *wire table* (see Figure 4.12c). Subsequently, each bus of an architecture graph provides such a wire table in its refinement. The specification of a bus wire table is defined by CCSL commands [Eis99].

To deal with the pins of a device each chip model uses a *pin manager*. Figure 4.12d) denotes the corresponding object-oriented modeling. A pin

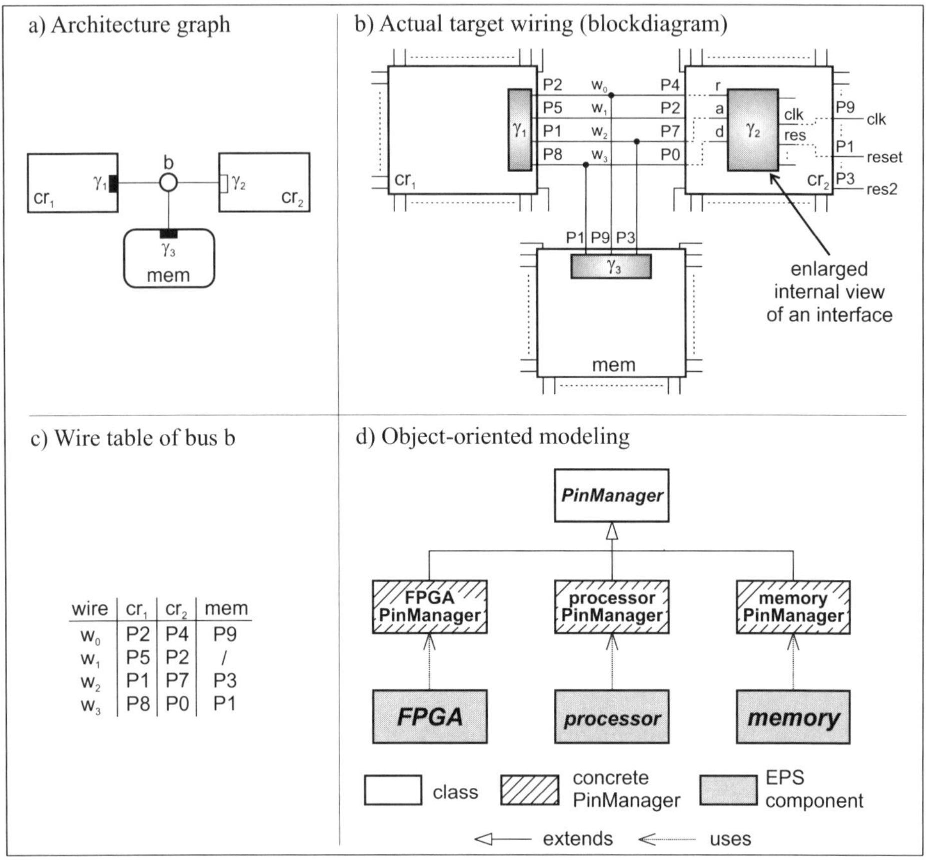

wire	cr_1	cr_2	mem
w_0	P2	P4	P9
w_1	P5	P2	/
w_2	P1	P7	P3
w_3	P8	P0	P1

Figure 4.12: Wiring architecture components.

manager is required to connect the device's internal circuits to pins (especially for FPGA computing resources). The available pins of a component are read from a database at instantiation of the component (see Prg. 3). The following issues have to be considered enabling automatic communication synthesis considering component wiring:

Wiring between components

The available wires and their connection to components are specified by a wire table. Subsequently, to establish a connection between components wires have to be allocated to channels to avoid short circuits and malfunction. In fact, the pin managers of the components have to reserve wires and assign them to channels.

Wiring within FPGA components

The generated interface circuits that are implemented by an FPGA computing resource have to be connected to device pins. For example, in Figure 4.12b) the port signal r of the generated interface circuit γ_2 is connected to device pin $P4$. The pin managers have to assure that internal and external wiring enables a correct connection between interface circuits.

Wire constraints

Wire constraints are necessary if the assignment of pins is not arbitrary. For example, a device's reset is usually connected to a dedicated pin. Each local entity (implementations of tasks, buffers and interface circuits) has to be connected to it. In Figure 4.12b) the external connected $reset$ signal has to be connected with the γ_2's res port signal.

The following example Prg. 6 shows the specification of the wire table for bus b (see Figure 4.12c) and demonstrates wire constraints concerning clk and $reset$ signals of the target shown in Figure 4.12b). Lines *2* ... *5* of Prg. 6 denote the first four columns of the wire table of bus b. For each wire there exists one such specification line. Lines *8* ... *10* specify the cr_2's device pins assigned to clock and reset. For example, line *9* specifies that the external signal $reset$ is connected to pin *P1* of cr_2. Lines *13* and *14* specify wire constraints applicable for the target. For example, each internal circuit using a reset has to be connected to a reset pin. Line *14* denotes the constraint where the internal port signal res of the interface γ_2 is connected to the reset and therefore to the external pin *P1*. Several clocks and resets can be used for a device. For example, cr_2 has an additional reset $res2$ that is connected to *P3* (see Line *10*).

Prg. 6: Specification of the wire table/constraints for Figure 4.12b) (language: CCSL)

```
1  % wire table of bus b
2  wire w0 b (cr1,P2) (cr2,P4) (mem,P9)
3  wire w1 b (cr1,P5) (cr2.P2)
4  wire w2 b (cr1,P1) (cr2,P7) (mem,P3)
5  wire w3 b (cr1,P8) (cr2,P0) (mem,P1)
6
7  % define clock and reset
8  assignCLK   clk   (cr2,P9)
9  assignRESET reset (cr2,P1)
10  assignRESET res2  (cr2,P3)
```

```

% wire constraints
assignPin cr2 clk clk
assignPin cr2 res reset
```

During the synthesis process the pin managers of the chip models assemble the information about the local assignment of device pins. For FPGA computing resources, the associated pin managers generate a constraint file denoting these assignments. These constraint files can be used by further compilation tools.

4.2.3 Automatic Communication Synthesis

The framework CCS provides automatic communication and interface synthesis by using a set of pluggable interface generators for the supported computing resources and memories (see Section 4.2.1). This section elaborates the embedding of interface generators into the object-oriented component model and outlines the process of device driver and interface circuitry generation.

Figure 4.13a) again visualizes the suggested chip model. It is mainly composed of a set of interface classes. Each interface class hosts a set of configurable interface generators. As an example, Figure 4.13b) shows the CSL architecture component. As modeled in Figure 4.10, it is an FPGA computing resource with an additional interface *PIO* (Parallel input/output). This interface has the two generators *PIO_8* and *PIO_16*. Essentially, these interface generators are able to generate appropriate interface circuitry for the *PIO* interface of the A7 device. Figure 4.13c) shows a refined class diagram of the *PIO* interface including these two interface generators *PIO_8* and *PIO_16*. The generators itself are derived from a class *InterfaceGenerator* providing the basic features of a generator. Figure 4.13d) visualizes the general template of an *interface generator*. During the synthesis process affected generators will be configured depending on the architecture component's embedding into the environment. Subsequently, they generates a set of code fragments. These code fragments may be written in different languages such as *VHDL* for R-type interfaces and *C* for P-type interfaces and have to be consistent with each other. They are automatically included into the final device's source code and provide the (hardware/software) interface for a point-to-point channel between communicating nodes. As a precondition, interface generators

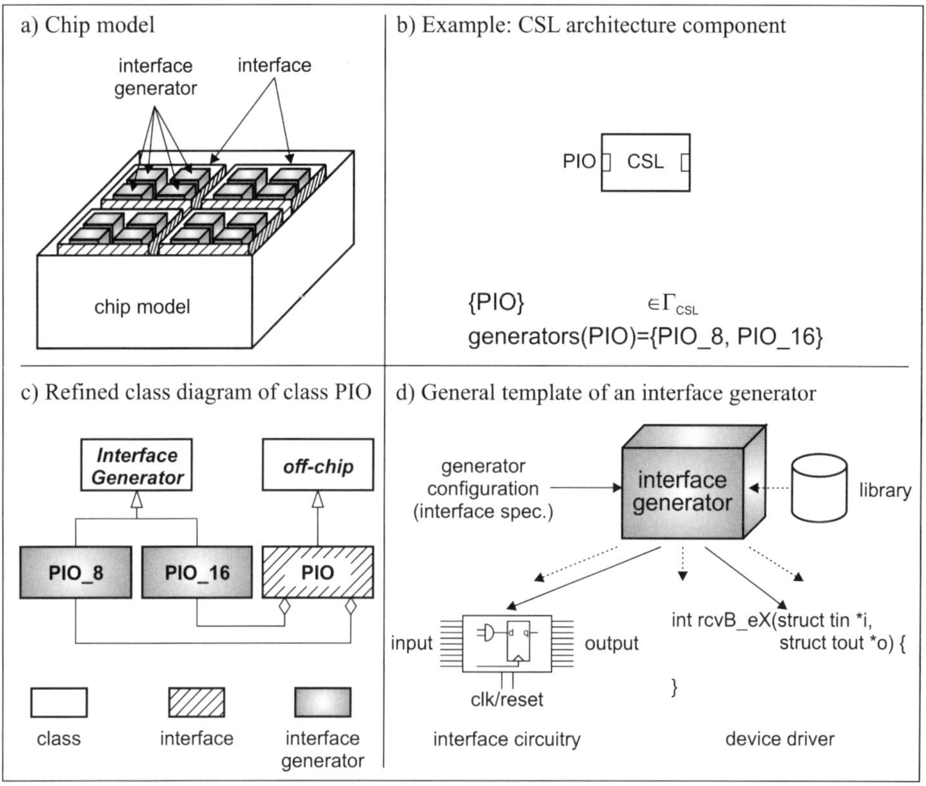

Figure 4.13: Embedding of interface generators.

base on the assumption that a point-to-point channel can be established using at most two communication nodes in a refined problem graph. This issue is covered by constraint *C3* in Section 3.2.3.

Cases of Interface Generation

The generation of the communication infrastructure is a rather challenging task, especially in the context of FPGA computing resources. This section focuses on the generation of interface circuitry and device drivers for the connection between two architecture components considering our object-oriented approach.

As an example, consider the following situation. An FPGA computing resource cr_1 is connected to a processor computing resource cr_2 using the interfaces γ_1 and γ_2, respectively (see Figure 4.14). This target architecture implements a refined problem graph *rPG* consisting of two communicating tasks x and y. The binding β shows the relation between the two graphs *rPG* and *AG*. The edge binding $\tau(e_1)$ and the interface generator selec-

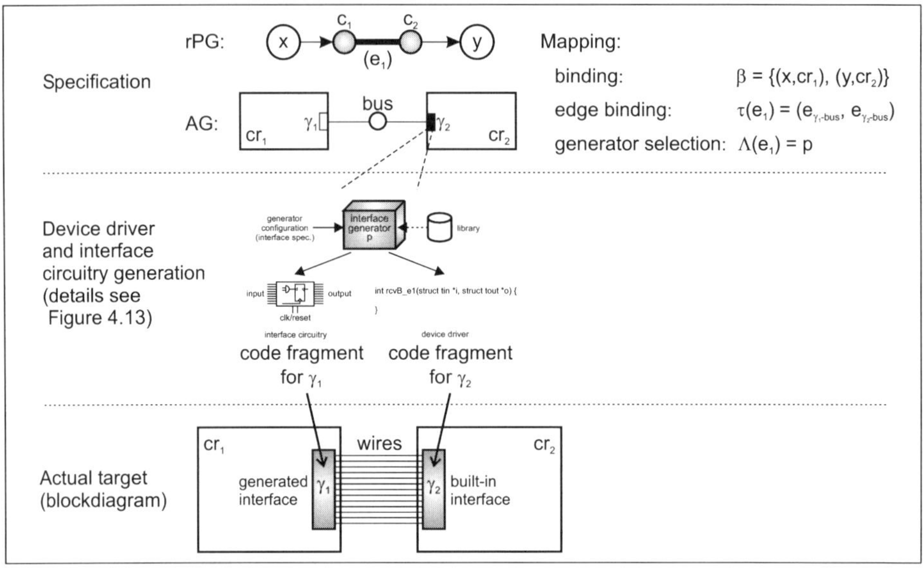

Figure 4.14: Generation of the communication infrastructure.

tion $\Lambda(e_1)$ denote the specification of the communication properties. To establish a channel between the two tasks, code fragments are necessary to implement the channel's access points c_1 and c_2. Our suggested approach resolves this situation as follows: Generally (as outlined in Section 4.2.1), an FPGA has no interface generator for counterpart interfaces. The only generator that is able to produce the necessary interface circuit for γ_1 is the interface generator p of interface γ_2. Subsequently, the interface generator p generates an interface circuit for γ_1 and as a counterpart, a corresponding device driver for γ_2. The generated code fragments for c_1 are exchanged via object interaction between the chip models.

Essentially, our suggested approach to interface synthesis distinguishes between five relevant cases of two connected architecture components (see Figure 4.15). The cases vary in the type of connected components as well as their interfaces. The following enumeration elaborates these five cases:

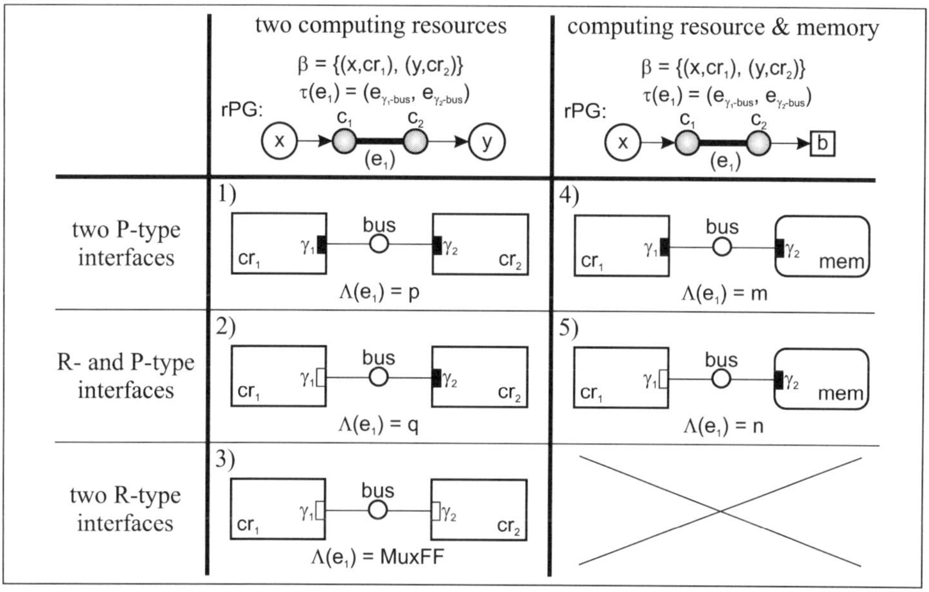

Figure 4.15: Cases for interface generation.

1) *Two connected computing resources with P-type interfaces*
This is the simplest case. Each of the interfaces γ_1 and γ_2 has a generator p that generates a device driver for the interface of its own component.

2) *Two connected computing resources with R- and P-type interfaces*
This case is generally known as *hardware/software interface* and has been elaborated in Figure 4.14. A further example is shown in Ex. 9.

3) *Two connected computing resources with R-type interfaces*
In general, a connection between such components (i.e., FPGAs) is not innately supported, i.e, no predefined interface circuit is available. However, it was shown that FPGA computing resources may have interface classes in their chip model (see Figure 4.10). In the considered case, the generator selection Λ denotes a generator *MuxFF* to be used. Therefore, each generator *MuxFF* produces an interface circuit for the interface of its own component.

4) *A computing resource with P-type interfaces connected to a memory*
A memory does not need a device driver. Therefore, no code has to be generated for interface γ_2. However, the memory component mem is "visible" from the computing resource cr_1 and is connected via memory mapped input/output or attached to a dedicated port. A memory component has a *MemoryManager* allocating and reserving the required address space for buffer b (not shown here). Considering constraint *C1*, the interfaces γ_1 and γ_2 have an instance of the generator m. Subsequently, the generator m of interface γ_1 is able to generate an appropriate device driver for interface γ_1 providing the selected buffer behavior. Required information from the memory component is exchanged via object interaction.

5) *A computing resource with R-type interfaces connected to a memory*
In general, the interface γ_1 has no appropriate interface generator providing a connection between cr_1 and mem (although it might have one depending on the modeling). In this case, the interface generator n of interface γ_2 produces the corresponding interface circuit for interface γ_1.

The above considerations can be expanded to a target architecture with several components connected to the same bus. In such a case, the involved chip models have to interact and jointly generate the interface circuits and device drivers. However, the affected interface generators have to support such an architecture.

4.2.4 Examples

This section presents examples concerning modeling and generating interfaces and device drivers.

Example 7 (Further chip models) *Figure 4.16 shows a set of additional chip models that have been used to establish target platforms [Tob99, HSS00, EPT99, EP00c, Kau00]. Again, these chip models emphasize the main advantages of the object-oriented approach namely:* simple retargeting, reuse, *and* simple device modeling. *For example, despite the fact that an FPGA computing resource has no predefined interfaces the* ALTERA *and* XILINX FPGA *chip models have been added an arbitrary number of* FPGABus8 *interfaces that provide channels between connected FPGAs of both FPGA families. In contrast, the interface* ReconfigBus *provides*

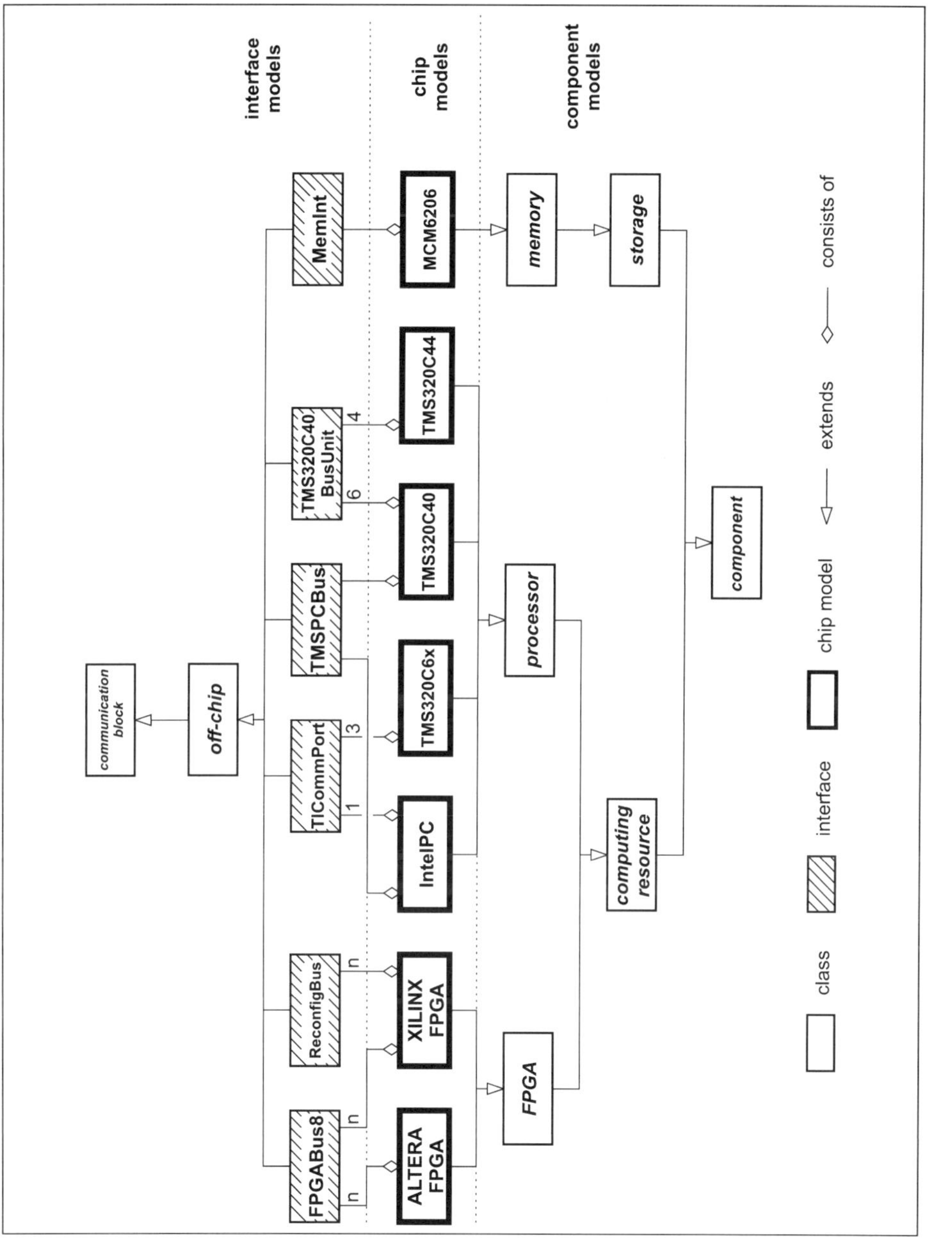

Figure 4.16: Class diagram of further chip models (UML notation [PJ99]).

a generator for interconfiguration communication between XILINX based FPGAs only.

Example 8 (FPGA on-chip communication channel) *This example outlines the implementation of a simple on-chip communication channel between tasks with compatible node ports (see Figure 4.17a). Figure 4.17b) shows the block diagram of the corresponding implementation. The two tasks are connected via a set of data and control wires. The output port of task x as well as the input port of task y consists of concurrent protocol FSMs and a register bank. Figure 4.17c) shows the two FSMs that are implemented in the input stage of task y.* FSM1 *implements a request/acknowledge protocol.* FSM2 *enables the register bank. To transmit a data item over the channel the signals* se1 *(send e1) and* re1 *(receive e1) have to be activated. After that, the FSMs of both tasks manage the proper data exchange. Subsequently, the* done *signals indicate the end of the transmission.*

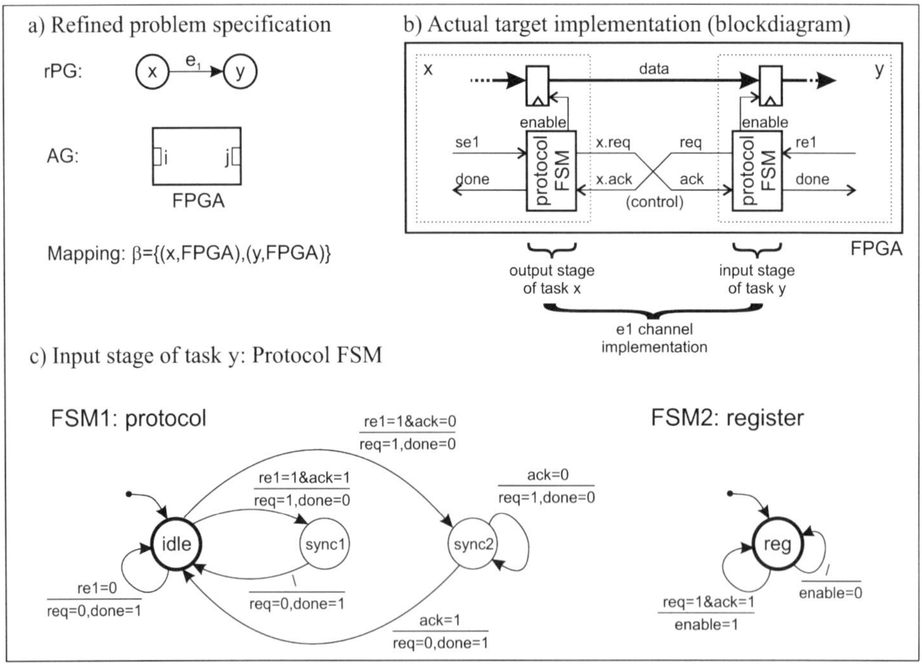

Figure 4.17: FPGA on-chip communication channel.

Example 9 (FPGA/microcontroller interface) *This example considers the structure and object-oriented modeling of a "hardware/software" interface between a XILINX XC4025 FPGA and a Motorola MC68340 microcontroller [ETT98]. The problem specification is given in Figure 4.18a). It is assumed that three channels between the FPGA and the microcontroller have to be implemented using a generator called* MotorolaStatusPort. *In our terminology, a status port provides memory mapped data communication between the two computing resources based on a* data port *and a* synchronization *port (see Figure 4.18b). Figure 4.18c) shows the block*

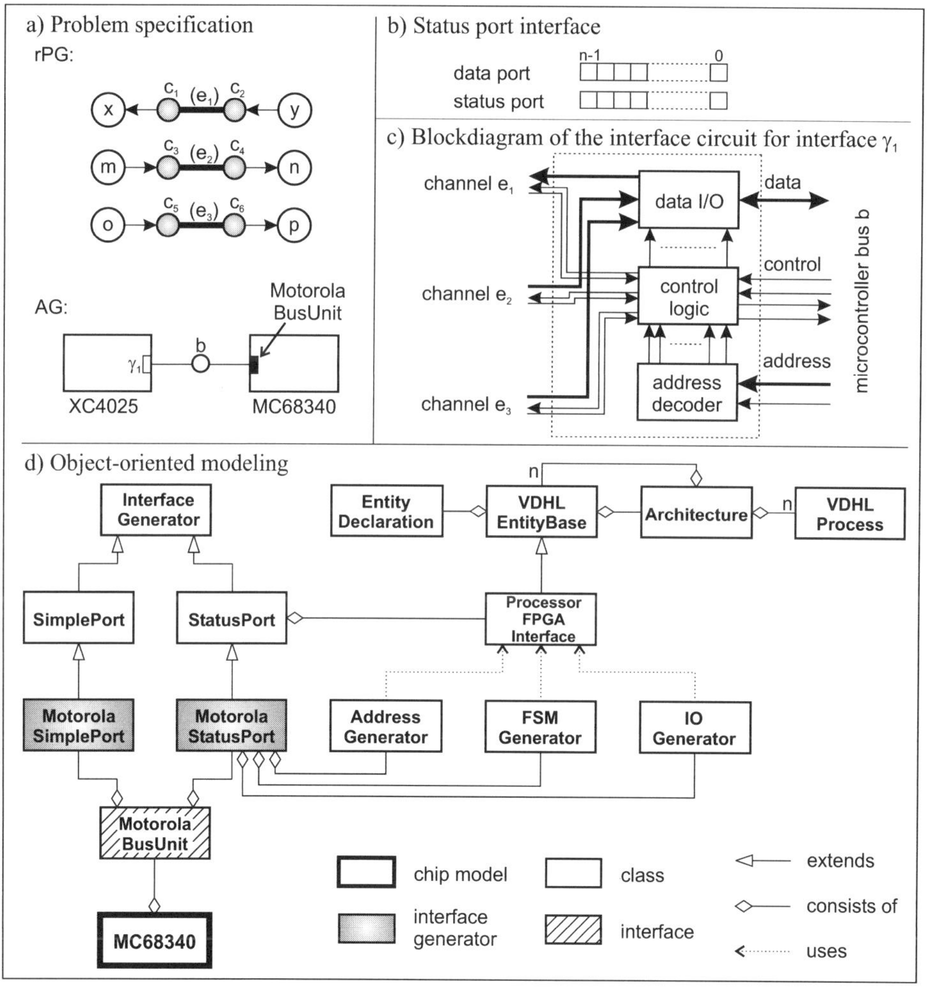

Figure 4.18: FPGA/microcontroller interface.

diagram of the interface circuit γ_1 *consisting of three parts (similar to signal groups on the bus): (i)* data I/O *providing intermediate registers between the channels* $e_1 ... e_3$ *and the microcontroller's data bus* b*, (ii) an address decoder detecting a microcontroller access in the memory region of the interface, and (iii) a control logic supervising the correct data transfer. The interface circuit is being generated by the interface generator of the microcontroller object. The generator MotorolaStatusPort is composed of tree generators reflecting the implementation. Each of them is responsible for a part of the interface. For example, the* FSM Generator *produces the* control logic *depending on the number of channels. Subsequently, the output of these generators is used to establish the interface circuit (*ProcessorFPGAInterface*).*

4.3 Problem graph

This section elaborates implementation issues of problem graph nodes on computing resources and memories considering channel access primitives.

4.3.1 Implementation Approaches

As outlined in Section 4.1.1, nodes use channel access primitives to send and receive data items. To discuss common implementation approaches for nodes and their related communication primitives, the problem specification of a task f in Figure 4.19a) is considered.

The assembling of access primitives and task implementation to a main program of a computing resource is usually performed using one of the following implementation schemes:

Direct insertion (inlining)
The access primitives are called within the task's source code (see Figure 4.19b) and are replaced by appropriate code fragments or function calls during compilation/linking time. This approach is quite often used during system development by hand if (i) an operating system for the computing resource and (ii) the sources of the tasks are available.

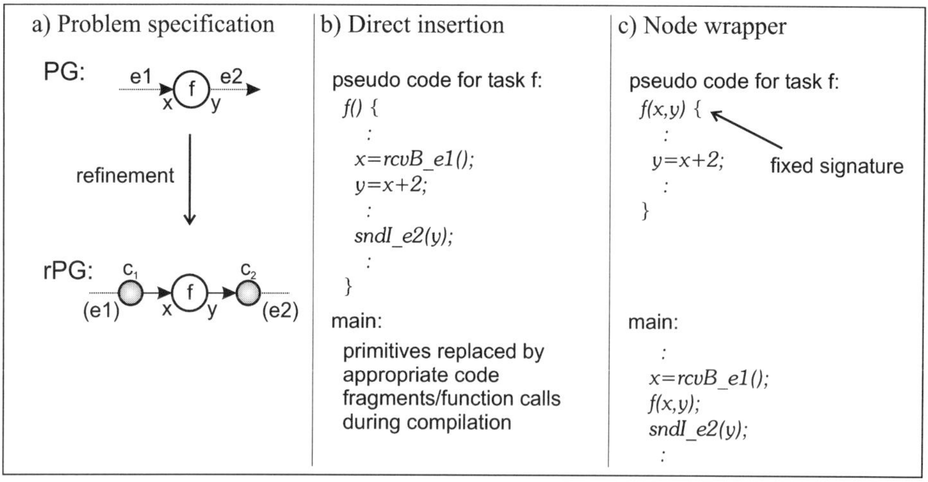

Figure 4.19: Implementation approaches.

Node wrapper
Generally, a wrapper encapsulates a dedicated functionality with its specific ports and protocols and provides the same functionality but with new ports and protocols [GHJV95]. Here, access primitives are wrapped around a task's core which is specified by a fixed signature (i.e., syntax description of the node ports in the implementation language) (see Figure 4.19c). Tasks are stored in a database in form of source code, object code, hard/soft macros, etc. This approach is used for embedded system implementation if (i) no operating system is available, and/or (ii) the source of the task is not available (which is often the case if IP cores are used).

For further investigations, the *node wrapper* approach is chosen for tasks and buffers due to its better suitability for automatic interface synthesis. Channel access primitives are being generated by the related object models of the components and are assembled to a main program for each computing resource during system synthesis.

4.3.2 Task

Figure 4.20 outlines a wrapper template of a task f on an FPGA computing resource with one input i, one output o, and one control input i_{ctrl} (for simplification only) [Eis96, ETT98]. The *core functionality* is sur-

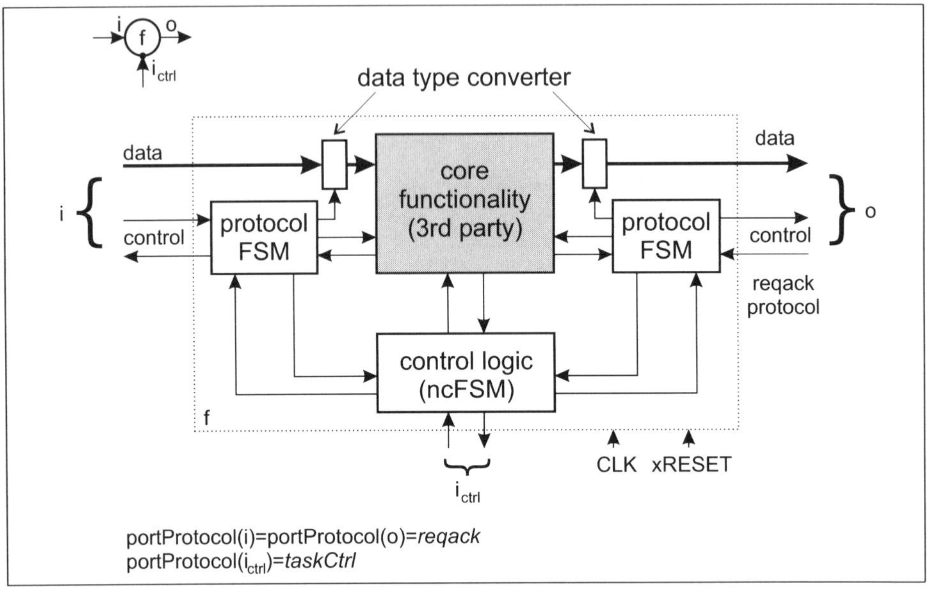

Figure 4.20: Wrapper template for a dedicated hardware core functionality.

rounded by small *protocol finite state machines* providing the adaptation from the core's protocols to the external protocols. *Data type converters* allow to adjust internal/external data type mismatches. The *control logic* (node control finite state machine *ncFSM*) orchestrates the local modules. This includes the core functionality and the protocol FSMs, and provides the control input port of the node i_{ctrl}. Prg. 7 describes a corresponding VHDL entity declaration (signature) outlining the control port i_{ctrl}, the input port i and the output port o.

Prg. 7: Entity declaration of a task f for a FPGA computing resource (language: VHDL)

```
entity f is
  port (
    CLK    : in  std_logic;                -- clock signal
    xRESET: in  std_logic;                 -- low active reset

    start : in  std_logic;                 -- control port
    done  : out std_logic;

    i_ack : in  std_logic;                 -- input port
     i_req : out std_logic;
     i     : in  std_logic_vector(31 downto 0);
```

```

    o_ack : in  std_logic;             -- output port
    o_req : out std_logic;
    o     : out std_logic_vector(31 downto 0)
  );
end f;
```

Figure 4.21 shows an example of a valid node control FSM for task f. It remains in the *idle* state until a $start$ message is issued to the control input port i_{ctrl}. This message is forwarded to all local connected FSMs. If the core functionality has come to an end ($doneCore$ issued) the ncFSM emits a $done$ message on the control port i_{ctrl}. The protocol on port i_{ctrl} is called *taskCtrl* protocol.

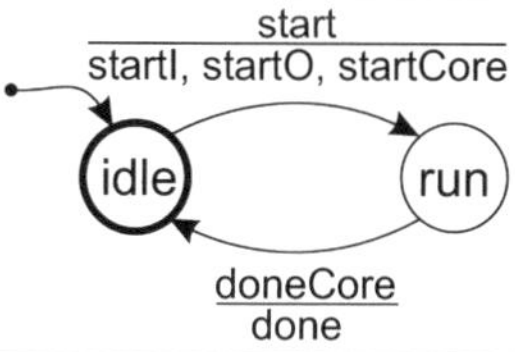

Figure 4.21: Example of a node control FSM.

The signature of a corresponding software task f with one input i and one output o is given in Prg. 8, line 2. Each task has an input data structure (e.g., f_in) denoting its input ports (lines 4 ... 8) and an output data structure (e.g., f_out) denoting its output ports (lines 10 ... 14). For each port, a *length* parameter specifies the number of data items to be read/write during one function call (see lines 7 and 13).

Prg. 8: Declaration of a software task f for a processor computing resource (language: C)

```
1  /* function signature */
2  int f(struct f_in *in, struct f_out *out);
3
4  /* input data structure */
5  static struct f_in {
6    int *i;
7    int i_length;
8  };
9
10 /* output data structure */
```

```
static struct f_out {
  int *o;
  int o_length;
};
```

Note that software tasks have no control input in their signature. Instead, the control input is indirectly represented by the activation of the task by the main program, or an associated dispatcher (e.g., see in Figure 4.19c).

4.3.3 Buffer

In [ETT98] a general template for a buffer (i.e., queue and register) has been suggested to be implemented on memories and FPGA/processor computing resources (see Figure 4.22). It consists of (i) an array of *memory cells* to store data items, (ii) read/write access functionality, and (iii) a

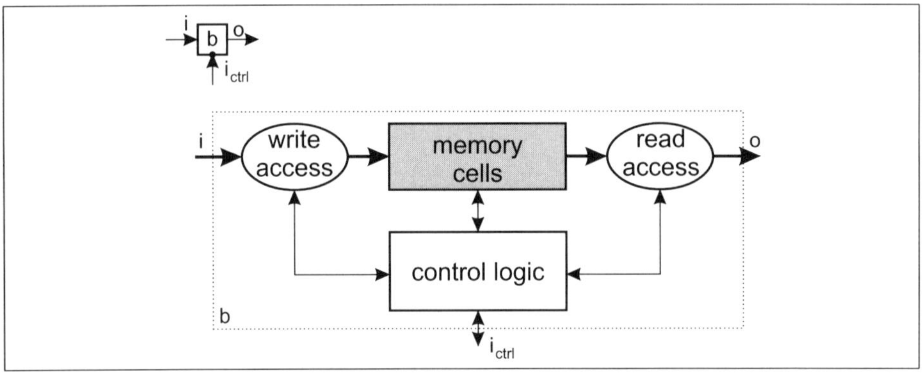

Figure 4.22: Implementation template of a buffer.

control logic. However, the actual implementation of a buffer can be quite different compared to a task's implementation and depends on the binding of the buffer. Generally, different (semiconductor) technologies are used to implement the memory cells and the rest of the template. For example, the write access as well as the control logic can be software functions, the memory cells can be located in a memory component, and the read access may be part of a hardware interface circuit. The following explanations focus mainly on the hardware implementation of a buffer. If applicable, additional comments are given concerning the software implementation.

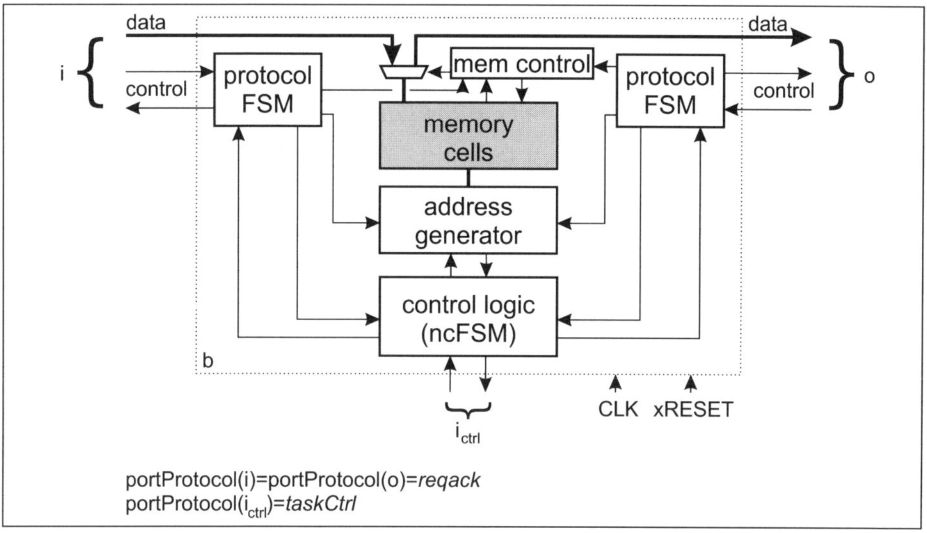

Figure 4.23: Template of a buffer in hardware.

In general, a buffer template for a hardware implementation is very similar to the task template (see Figure 4.23). It consists of (i) an array of *memory cells*, (ii) two *protocol FSMs* for protocol adaptation, (iii) an *address generator* defining the access's memory cell, (iv) a memory control logic *mem control*, and (v) the node's *control logic*.

Buffer implementation
To elaborate a buffer's hardware implementation the same problem specification used to establish the taxonomy of communication types is considered (see Figure 4.7a). Depending on the binding of a buffer four significant cases arise for buffer implementation:

a) $\beta = \{(x, cr_1), (b, cr_1), (y, cr_1)\}$*: no buffer splitting*
This is the simplest case concerning automatic synthesis. The buffer is implemented using the suggested template taken from a library (see Figure 4.24a1). No communication node is necessary in the refined problem graph and no dedicated interface circuitry is required within the implementation. The node implementations are directly connected using an on-chip communication channel. Simultaneous read/write accesses are resolved by the buffer implementation. However, this case can be rather area inefficient. Namely, memory cells are implemented either using (i) configurable

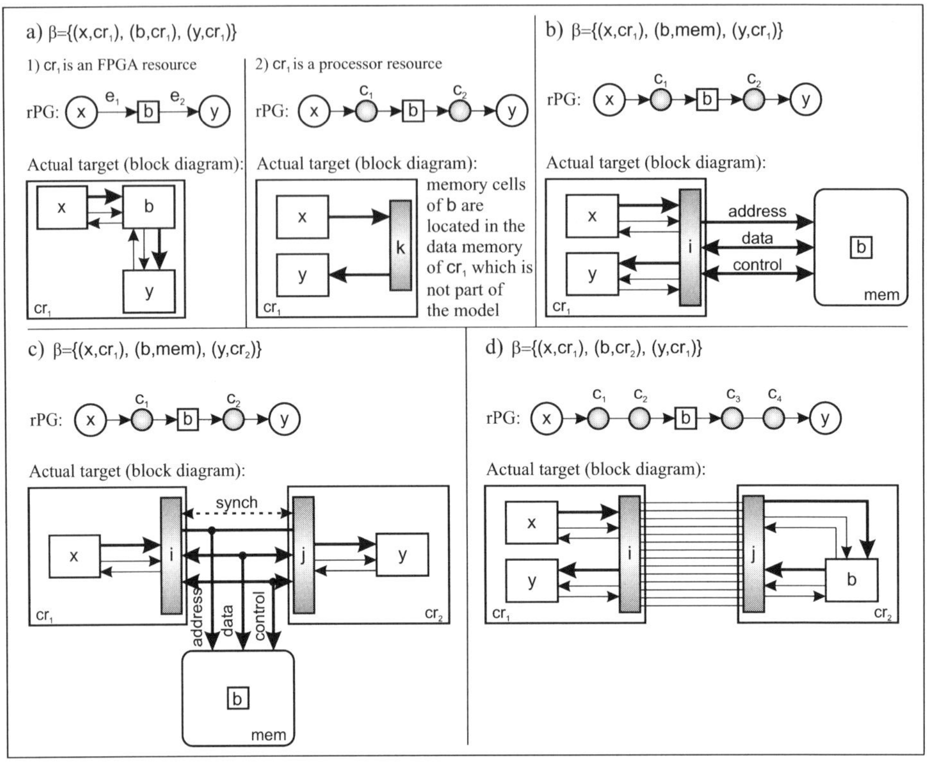

Figure 4.24: Refined problem graphs.

logic cells (low memory density), (ii) configuration memory of configurable logic cells (medium memory density), or (iii) integrated memory as provided by certain FPGA families (high memory density). If cr_1 is a processor computing resource the memory cells are automatically located in the associated data memory of cr_1 (high memory density). In this case, communication nodes are necessary and represent the read/write access primitives to the memory cells (see Figure 4.24a2).

b) $\beta = \{(x, cr_1), (b, mem), (y, cr_1)\}$: *distributed buffer implementation*

In this case, the memory cells are located in the memory component mem and exactly one computing resource (here cr_1) accesses the buffer (see Figure 4.24b). Therefore, the template is *divided into two parts*; one part implements the memory cells and the other part implements the other parts of the template within the computing resource. The communication nodes c_1 and c_2 represent the interface to the buffer and provide the write and read access to the memory cells. Simultaneous read/write accesses are resolved by the interface implementation. If cr_1 is a processor computing resource the implementation is almost equal to case a). Only the location of the memory cells is different.

c) $\beta = \{(x, cr_1), (b, mem), (y, cr_2)\}$*: distributed buffer implementation*
This kind of buffer implementation often appears in heterogeneous embedded systems [OLB98]. Tasks of two connected computing resources communicate using a shared memory region. Here, the implementation of the buffer template comprises *at least three parts* implemented by the involved architecture components. The memory cells are located in the memory component mem and each of the tasks x and y uses its own interface circuit/device driver to access the memory cells (see Figure 4.24c). However, due to the distributed buffer implementation simultaneous accesses to the memory cells have to be *synchronized* among the involved interfaces and device drivers.

d) $\beta(x) = cr_1$, $\beta(b) = cr_2$, $\beta(y) = cr_1$*: no buffer splitting*
As outlined in Figure 4.24d) the buffer template is implemented without any splitting and is very similar to case a). Here, two communication nodes are required for each edge.

As an example, Figure 4.25 shows the block diagram of an SRAM interface circuit providing a set of $^n/_2$ FIFO buffers. It enables the communication between a set of tasks bound to an FPGA computing resource and a set of buffers bound to a connected MCM6206 SRAM [Tob99, EP00c]. As outlined in the buffer template (see Figure 4.23) each of the n competing channels consists of a small protocol FSM and an address generator logic. These two stages together either represent a write or a read access functionality of a single buffer. The block *mem control/ncFSM* provides the necessary access control to the memory cells, e.g., by a fixed priority scheme.

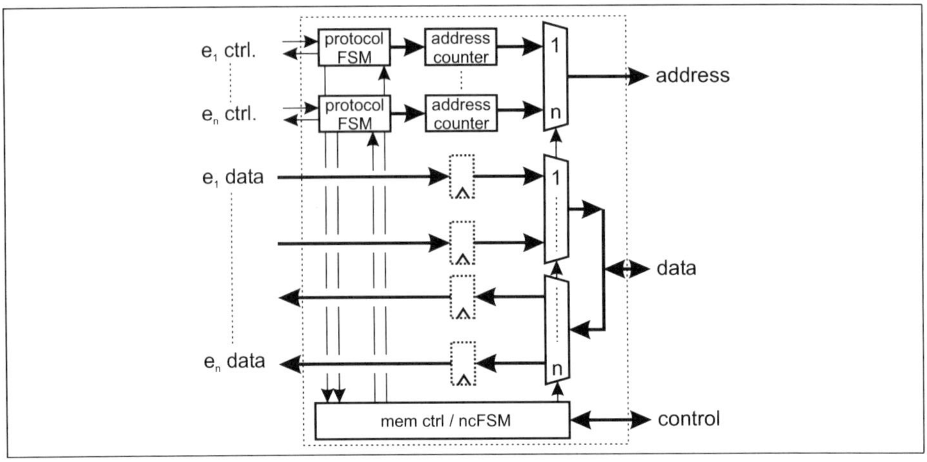

Figure 4.25: Block diagram of an SRAM interface.

4.3.4 Dispatcher

Each computing resource requires a *"main function"* describing the resource's execution. For an FPGA resource, a structural representation of connected VHDL *entities* is necessary. Here, entities describe tasks, buffers, dispatchers, as well as interface circuitry. Processor computing resources have a *main()* function that executes a set of *C* functions. These functions include tasks and channel access primitives. Both computing resource types require at least one dispatcher (see constraint *C4* in Section 3.2.3). It implements a static or dynamic schedule that is captured by the dispatcher finite state machine. Depending on the type of computing resource a dispatcher is bound to, three different kind of dispatcher implementations emerge (see Figure 4.26):

a) *FPGA computing resource*
A dispatcher bound to an FPGA computing resource is implemented as VHDL entity whose outputs are connected to control inputs of other nodes. The implementation has the same internal structure and entity declaration like a hardware task implementation. As an FPGA resource enables parallel node implementations several dispatchers can exist concurrently.

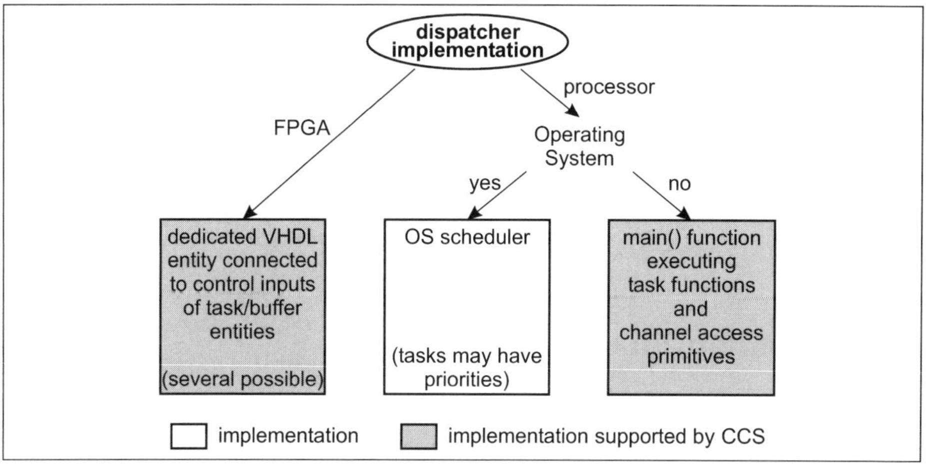

Figure 4.26: Implementation approaches for dispatchers.

b) *Processor computing resource with operating system*
In this case, the dispatcher models part of the OS scheduler. Tasks and buffers are activated/suspended depending on their access to blocking and non-blocking channel access primitives. Furthermore, tasks may have priorities to enable the correct activation sequence. However, this kind of implementation was not focus of this thesis.

c) *Processor computing resource without operating system*
The dispatcher's finite state machine is part of the main() function. It repeatedly executes sequences of task and channel access primitive function calls and implements a static or dynamic task schedule [ET98a, EZT00].

Ideally, a dispatcher implementation is being generated based on (i) scheduling information of tasks and (ii) the semantics of channel access primitives. However, it was not a major focus of the thesis to generate dispatchers. Therefore, although a *"main function"* is generated for each computing resource extra work is required to implement a dispatcher's behavior.

4.4 Reconfigurable Systems

The refinements concerning the EPS model are valid for the RPS model as well. Additionally, an RPS implementation has to provide the dynamic reconfiguration of FPGA computing resources which involves the implementation of configurators. This section elaborates the implementation issues of configurator nodes.

4.4.1 Configurator

A configurator "supervises" the scheduling of configurations for one or several FPGA computing resources (see Section 3.3.1). Depending on the type of computing resource a configurator is bound to, two different kind of configurator implementations arise (see Figure 4.27):

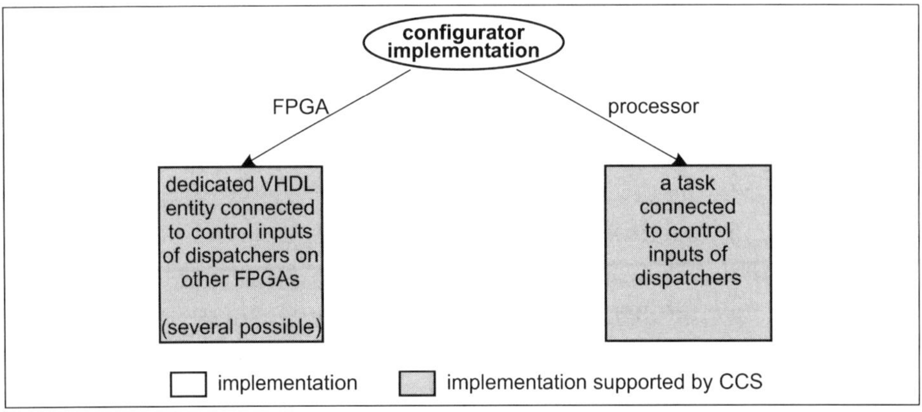

Figure 4.27: Implementation approaches for configurators.

a) *FPGA computing resource*
Similar to a dispatcher's implementation a configurator on an FPGA computing resource is implemented as VHDL entity. Its outputs are connected to control input ports of dispatchers on other FPGAs. Its implementation has the same internal structure and entity declaration like a hardware task's implementation. Several configurators may exist concurrently.

b) *Processor computing resource*
A configurator on a processor computing resource is implemented like another task that reads/writes channels via channel access primitives. It is independent on the presence/absence of an operating system.

Ideally, a configurator's implementation is being generated based on scheduling information of the supervised configurations. However, as it was not a major focus of the thesis to generate configurators, manual extra work is required to implement the configuration schedule sequence for the configurators.

4.5 Summary

This chapter elaborates refinements and implementation issues of the proposed EPS and RPS models.

By introducing communication nodes into a problem graph, edges are refined to model communication channels. These communication nodes represent device drivers and interface circuits and provide *blocking* or *non-blocking* semantic. Based on different use cases various edge refinement strategies are discussed leading to our proposed *taxonomy of communication types*. Depending on the binding of nodes of the problem graph, four implementation types arise for communication channels: *on-chip*, *off-chip*, *interconfiguration*, and *interconfiguration communication with an open channel*.

To establish an appropriate communication infrastructure an object-oriented approach is proposed. For each component of the architecture graph there exists a corresponding software object (*chip model*) modeling the properties of the actual device. These objects are composed of *configurable interface generators* that are able to generate the specified communication infrastructure. Comparing to traditional approaches such as the use of communication libraries our approach has the advantage of

- a simple retargeting of the interface generation,
- reuse of the generators,
- simple modeling of a device, and
- object interaction providing simple synthesis algorithms.

The proposed approach includes modeling of the actual *component wiring* enabling the automatic assignment of device pins to communication channels.

Based on a *wrapper* approach the implementation issues are discussed concerning the embedding of communication primitives and node implementations into a *"main program"* for processor and FPGA computing resources. This includes the joint interaction of dispatchers and configurators to establish the overall schedule of the specified system.

Chapter 5

Optimization and Synthesis

This chapter gives a coherent view between the proposed models (see Section 3), their refinement (Section 4), and a corresponding synthesis flow. At the beginning, the synthesis flow is elaborated (Section 5.1). It covers the path from a problem specification down to an implementation. The next Section 5.2 outlines an optimization methodology to reduce the number of problem graph nodes and required device pins. In Section 5.3, a framework is presented that supports the proposed synthesis flow [EP00b]. Section 5.4 outlines an extended example for the specification, refinement, and synthesis of a simple SDF graph. Finally, Section 5.5 provides a short summary of the chapter.

5.1 Synthesis Flow

This section outlines a synthesis flow based on our proposed models and refinements. It starts by specifying a problem and ends with a corresponding implementation. Figure 5.1 briefly outlines the proposed synthesis flow that consists of three phases and comprise nine major steps:

Specification
During the specification phase a synthesis problem is captured and described by using the proposed formalisms EPS and RPS, respectively.

1) *Problem capturing*
The first step captures the considered problem using the EPS formalism. Therefore, a problem graph, an architecture graph, as well as an appropri-

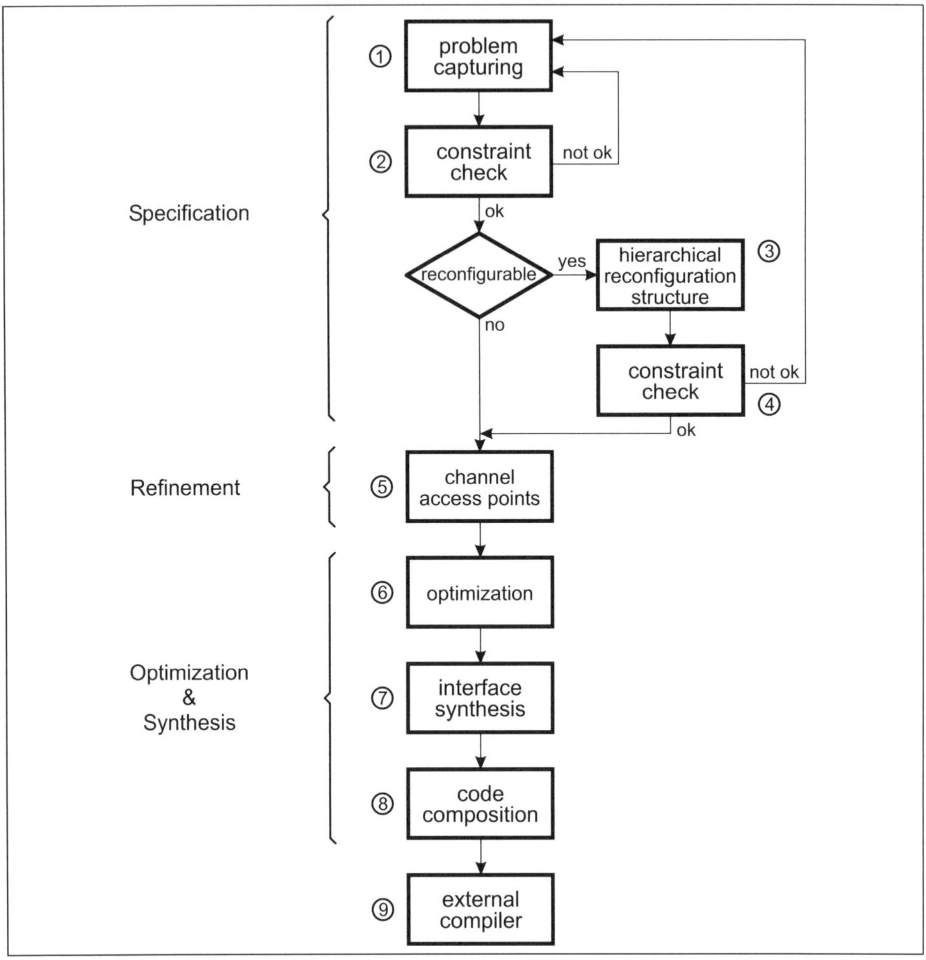

Figure 5.1: Simplified synthesis flow.

ate mapping has to be defined. Additionally, at least one dispatcher node per computing resource is necessary to models the activation of problem graph nodes.

2) *Constraint check*
For a feasible specification, it has to be assured that the constraints *C1* ... *C6* are met (see Section 3.1.2 and Section 3.2.3). This may involve the insertion of dispatcher as well as routing nodes into the problem graph.

3) *Hierarchical reconfiguration*
In case of dynamic reconfigured FPGA components the RPS formalism enables to describe the necessary hierarchical reconfiguration structure (see Section 3.3). Here, the mapping covers four steps where the node-configuration assignment and the configuration-component assignment determine the dynamic problem graph parts. Furthermore, at least one configurator node on a host, as well as one dispatcher node per configuration has to be inserted (see Section 3.3.3).

4) *Constraint check*
Besides the constraints of step *2* a feasible mapping satisfies constraints *C7* ... *C9* (see Section 3.3.3). Furthermore, depending on the application scenario additional constraints *C10* ... *C16* have to be satisfied (see Section 3.3.5 ... Section 3.3.7).

Refinement
At this point, the problem specification is finished and the communication infrastructure has to be established.

5) *Channel access points*
Based on the edge binding (see Section 3.2.3) channel access points are inserted for the edges of the problem graph (see Section 4.1.1). These access points are related with component interfaces and enable the generation of the communication infrastructure.

Optimization and synthesis
The last phase optimizes the specification and generates the specified interface and device driver codes, as well as code for each computing resource.

6) *Optimization*
Here, optimization methodologies are applied (see Section 5.2). The goal is to reduce the number of problem graph nodes, as well as to reduce the required device pins to implement the communications channels.

7) *Interface synthesis*
For each channel access point interface circuitry or device driver needs to be generated (see Section 4.2.3). The generation methodology bases on our proposed object-oriented component model.

8) *Code composition*
The generated code fragments for the channel access points as well as codes for the node's implementation are assembled for each configuration and computing resource. This may include the generation of *command scripts* for the following compilation step.

9) *External compiler*
The above steps produce source code for each computing resource and configuration. Subsequently, these codes have to be compiled by the use of appropriate compiler tools for each computing resource.

5.2 Optimization by Object Sharing

The following optimization methodology is applicable to RPS specifications in order to reduce (i) the number of problem graph nodes, (ii) the number of necessary device pins, and (iii) the area of interface circuitry [EP00b].

Problem description
Figure 5.2 shows an architecture graph AG with a host and two FPGAs *fpga1* and *fpga2*. *fpga1* has two configurations δ_{11} and δ_{12}. *fpga2* implements three configurations δ_{21}, δ_{22}, and δ_{23}. For simplification, only the configurator node ρ as well as the dispatcher nodes (e.g., σ_{22}) of the problem graph PG are shown. At first sight, six routing nodes ($r_{111} \ldots r_{123}$) are necessary. As a consequence, interface circuitry and device pins for 14 channels on *fpga1* and 6 channels on *fpga2* would be required.

Nevertheless, the additional routing nodes as well as the number of device pins can be reduced by applying *optimization by object sharing*.

Solution
The proposed optimization methodology makes use of the fact that some of the configurations are exclusive. Objects that can be shared comprise problem graph nodes and device pins. Although only routing nodes have been considered, object sharing could be applied to the original nodes of the problem graph as well. Nodes can be divided into three groups that demand for different interface circuitry (see Figure 5.3). Nodes of type a are connected only to nodes within the same configuration and are thus not

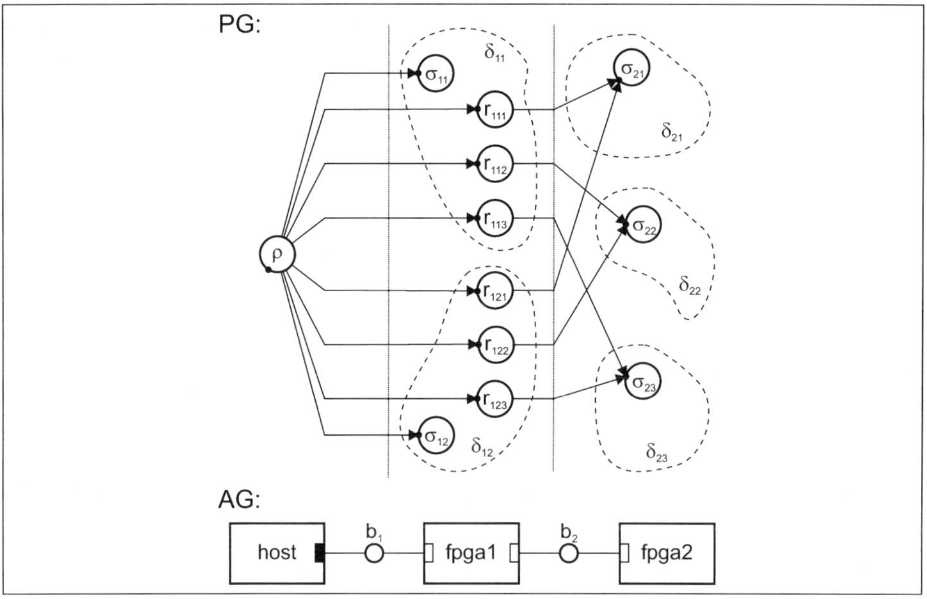

Figure 5.2: Routing node distribution.

amenable to the proposed optimizations. Nodes of type b are connected to nodes in the same configuration as well as to nodes bound to other architecture components. Nodes of type c have only connections to nodes bound to other architecture components. Connecting nodes of type b and c requires device pins and is the main target of object sharing. The routing nodes, which are of type c, benefit most from object sharing optimizations. The proposed optimizations are split into two groups:

- *Pin sharing:* Several communication channels alternately use the same set of device pins.
- *Node sharing:* The implementation of a node is shared between several communication channels in one configuration.

Pin sharing

The wiring of architecture graph components requires a set of device pins. Each of these pins is associated with a channel to enable data communication (see Section 4.2.2). But, each device has only limited pin resources which can be overcome by sharing pins among several channels.

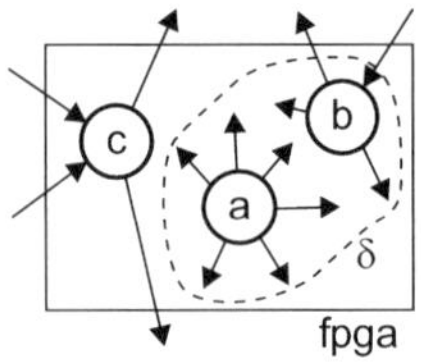

Figure 5.3: Node types: a, b, and c.

Subsequently, pin sharing applies to certain edges of the node types b and c. In the Virtual Wire project pin sharing has been used for compile-time reconfigurable resources only [BTD+97, BTA93]. Here, pin sharing has been extended to time-exclusive configurations of run-time reconfigurable resources. Therefore, two types of pin sharing are differentiated:

- *Intraconfiguration* pin sharing allows communication channels in the same configuration to share device pins by adding multiplexing and demultiplexing interface circuitry [BTA93]. This technique can be applied to both CTR and RTR systems.

- *Interconfiguration* pin sharing allows communication channels in different configurations of the same FPGA to share device pins. The channels would otherwise be mapped to different pins as configurations of other FPGAs could require that both channels exist at the same time. Interconfiguration pin sharing applies only to RTR systems.

As an example, Figure 5.4a) shows part of the problem graph of Figure 5.2. Applying intraconfiguration pin sharing to edges $\rho - r_{111}$, $\rho - r_{112}$, and $\rho - r_{113}$ implies multiplexing/demultiplexing interface circuitry for the channel access points c_3 and c_4 (see Figure 5.4b).

As an example for interconfiguration pin sharing, Figure 5.5a) shows again a part of the problem graph of Figure 5.2. The configurator controls the dispatchers via the edges $\rho - d_{11}$ and $\rho - d_{12}$. As the two dispatchers reside in exclusive configurations, the corresponding pins of the two edges can be shared. In Figure 5.5b) this is expressed by the channel access points c_1 and c_2.

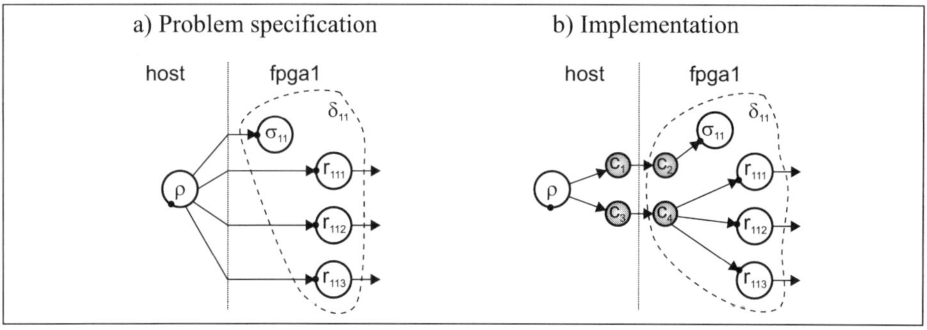

Figure 5.4: Intraconfiguration pin sharing.

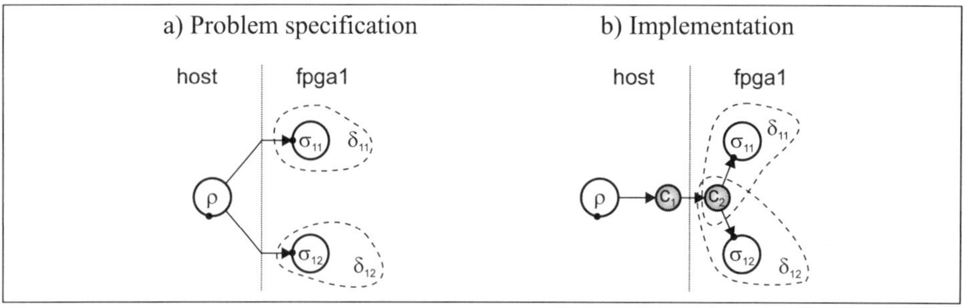

Figure 5.5: Interconfiguration pin sharing.

Node sharing

Node sharing is a methodology where several problem graph nodes assigned to the same configuration share a single physical implementation on the target architecture. Generally, node sharing can be applied to all three node types a, b, and c. Node sharing to original problem graph nodes allows to trade-off between required FPGA area and execution time and could require additional buffers. Node sharing for original problem graph nodes actually means to modify the mapping determined by front-end tools. Hence, node sharing is considered for routing nodes only (nodes of type c), which is extremely valuable if combined with pin sharing.

As an example, Figure 5.6a) shows a problem specification including *fpga1* with one configuration and *fpga2* with three configurations. The communication channels between configurator ρ and dispatchers $d_{21} \ldots d_{22}$ are routed over *fpga1* and require routing nodes $r_{111} \ldots r_{123}$. Obviously, these nodes can share one physical implementation as the three configurations of *fpga2* are exclusive. Furthermore, the sharing of nodes can be combined with intraconfiguration pin sharing of the three channels between the host and *fpga1* and interconfiguration pin sharing between *fpga1* and *fpga2* (see Figure 5.6b).

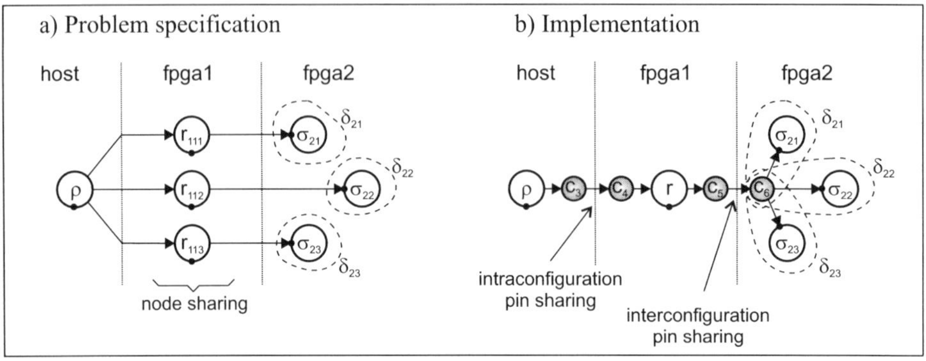

Figure 5.6: Node and pin sharing.

Finally, Figure 5.7 shows the optimized implementation for Figure 5.2 after applying node and pin sharing. The hierarchical reconfiguration structure consists of a configurator node on the host, one dispatcher node per FPGA configuration, and two routing nodes on *fpga1*. Overall, three pin sets are required on *fpga1* and one set of pins on FPGA *fpga2* to implement the reconfiguration structure.

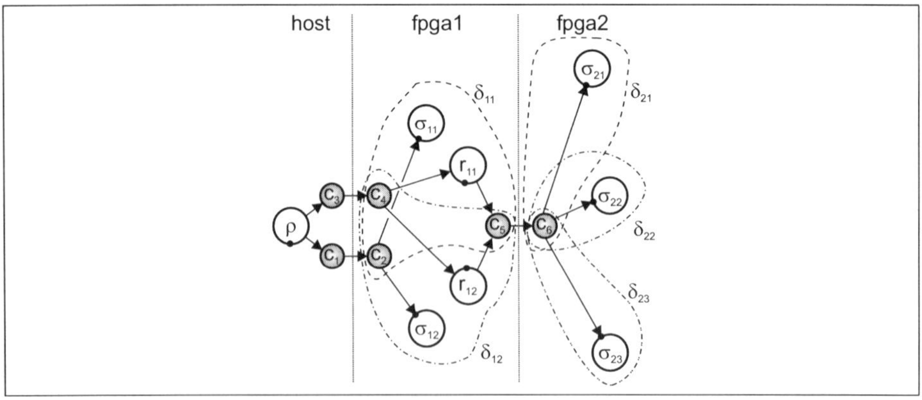

Figure 5.7: Object sharing optimizations.

5.3 CCS framework

The initial goal of the CCS (*C*ommunication *C*hannel *S*ynthesis) framework was to establish a "thin thread" from a problem specification down to an implementation to validate the proposed approaches. However, it evolved to a test platform that allows to plug-in design and implementation tools to generate a communication infrastructure for problem specifications based on the GPS formalism [EP02]. During the thesis, it helped (i) to get experience in interface and device driver synthesis, (ii) to develop the object-oriented component model and corresponding synthesis algorithms, and (iii) to establish a synthesis flow for embedded (reconfigurable) systems.

This section gives a brief overview about the framework as well as its embedding into existing co-design frameworks.

5.3.1 Overview

The structure of the framework is outlined in Figure 5.8. Shaded boxes denote areas of thesis contributions. The *data object repository* forms the heart of the framework. It captures design data which is stepwise refined from the initial problem specification down to the final implementation. Tools that generate, modify, and refine repository objects are divided into *front-end*, *back-end*, and *off-line* tools.

Front-end tools

Basically, front-end tools are used to establish the problem specification in the sense of Section 3. *Design capture* tools allow to formulate a design problem as data object comprising a problem graph (e.g. [Jan00]). *Architecture synthesis* tools extend this data object by adding (i) an appropriate architecture graph, (ii) a mapping between problem and architecture graph, and (iii) information about the execution order of problem graph nodes and configurations (e.g., [PB99]). These steps are based on primary *estimations* about the data objects. Furthermore, *simulation* and *verification* tools may be applied to such problem specifications.

Back-end tools

Back-end tools include problem refinement (see Section4), algorithms for synthesis, as well as tools to implement a certain synthesis design flow

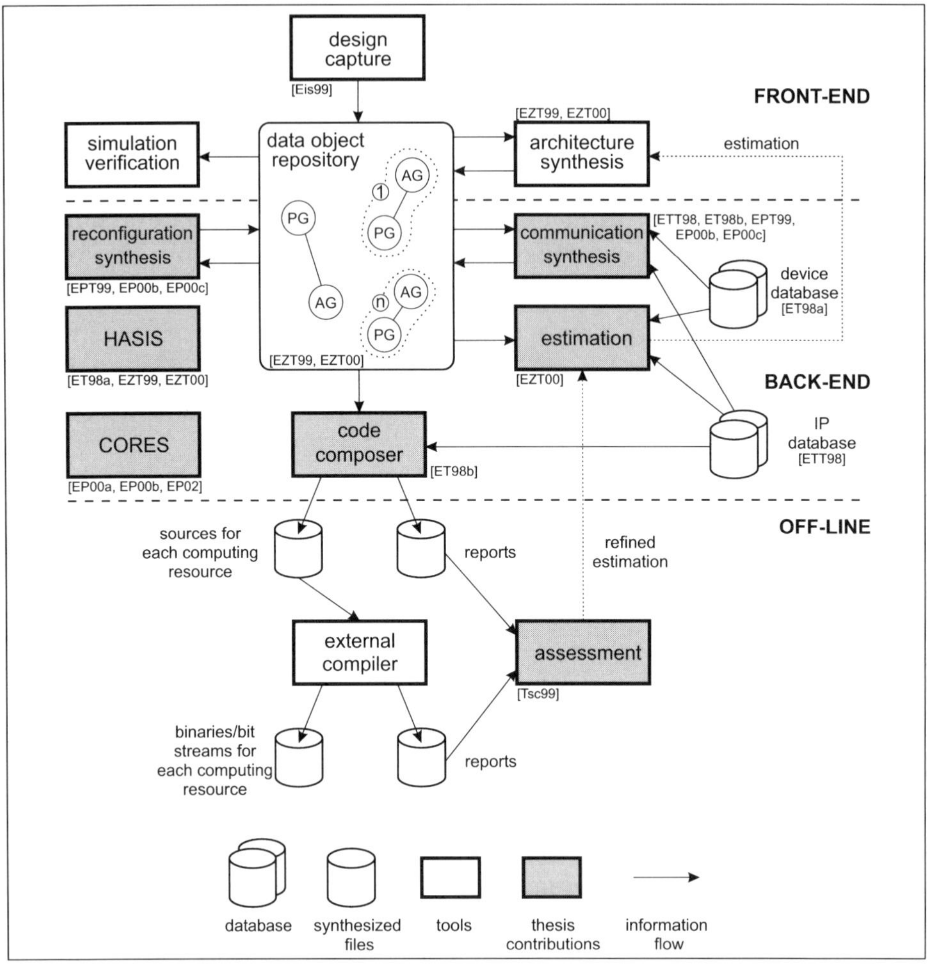

Figure 5.8: Overview about the structure of the CCS framework.

(see Section 5.1). Two databases support these tools: the *device database* provides chip models for components of the architecture graph, and the *IP database* stores problem graph node implementations, i.e., IP (intellectual property) cores. In both databases, each entry is accompanied with a description of its most important features such as port descriptions, timing/area estimation, and required information for synthesis. A first tool is *reconfiguration synthesis* which refines a specification by introducing a hierarchical reconfiguration scheme. As a result, each configuration comprises the specified tasks and buffers as well as additional tasks responsible for reconfiguration control. *Communication synthesis* tools establish communication channels between communicating problem graph nodes by automatically generating interface circuitry and device drivers that are modeled by channel access points. *Estimation* tools assess refined data objects and report on requested design parameters. These estimations drive front-end tools in a next design iteration. One example is the overhead in terms of FPGA area required for implementing the hierarchical reconfiguration scheme. *Code composer* tools assemble IP cores, reconfiguration scheme, as well as the generated interface circuitry and device drivers for each configuration and computing resource. The two remaining tools *HASIS* (*H*ardware/*S*oftware *I*nterface *S*ynthesis) and *CORES* (*C*onfigurator for *R*econfigurable *E*mbedded *S*ystems) provide the design flows for static reconfigured architectures and dynamic reconfigured architectures by activating the required tools and algorithms.

Off-line tools
The code composer generates source files only. Subsequently, *external compiler* tools, such as an FPGA compiler, are necessary to translate the produced sources into binaries and bit streams for each computing resource. Furthermore, *assessment* tools (e.g., [Tsc99]) are able to provide a rating of the generated sources. This enables a refinement of the primary estimations, and a refinement of the elements stored within the databases.

5.3.2 Embedding the CCS framework

The CCS environment may serve as a backend synthesis framework for a model based co-design flow. Figure 5.9 shows an appropriate layered view of the embedded framework.

The lowest layer consists of the CCS tools discussed in the last section (hatched area). On top of that, a *JAVA API* provides (i) capturing the problem specification, (ii) controlling the synthesis process, and (iii) enabling access to estimation tools and the databases. Subsequently, this API qualifies a seamless framework integration into an existing JAVA environment that covers *problem analysis*, *designspace exploration*, and *design of a target platform*. As an example, in [Tch98, EZT99] a corresponding case study has been implemented using (i) the MOSES environment as a graphical user interface (*GUI*) for problem specification [Jan00], (ii) an evolutionary algorithm for architecture synthesis [ZT99], and (iii) CCS as back-end synthesis tool. Furthermore, a command line shell CCSL

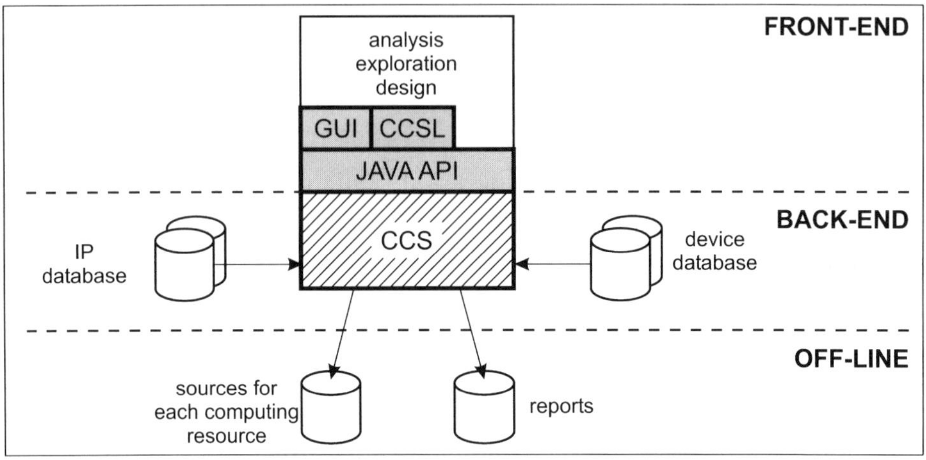

Figure 5.9: Embedding the CCS framework.

(*C*ommunication *C*hannel *S*ynthesis *L*anguage) [Eis99] provides an interactive interpreter that translates between the JAVA API and a pure ASCII interface.

5.4 Extended example

In this example, specification, refinement, and synthesis of a simple SDF graph is considered (see Figure 5.10). Thereby, the synthesis flow as described in Section 5.1 is applied. In the sense of this thesis, the focus of interest is on the overhead caused by the communication infrastructure which is necessary to enable the communication.

The following discussions show, that the proposed models and methodologies are useful for tools seeking for optimal system implementations [EZT00]. Such tools considerably depend on appropriate system descriptions that enable an efficient search of alternative implementations.

5.4.1 Problem Specification

The target is described by the *architecture graph* in Figure 5.11. It consists of a $host$ computing resource which is able to reconfigure the two connected FGPA computing resources $fpga1$ and $fpga2$. Furthermore, memories $mem1 \ldots mem3$ are connected to the FPGA resources. As an exam-

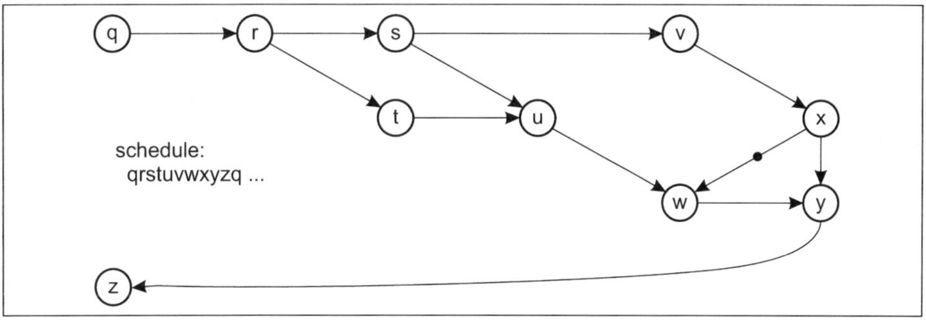

Figure 5.10: Simple SDF (Synchronous Dataflow) graph.

ple, an RPS specification of the SDF graph is elaborated, where both FPGAs are dynamically reconfigured during the application's run-time. An initial *problem graph PG* as well as a potential binding β^* is specified in Figure 5.12a). It may have been generated by automated front-end tools or manually. Here, for each edge of the SDF graph a buffer node has been inserted to store intermediate data.

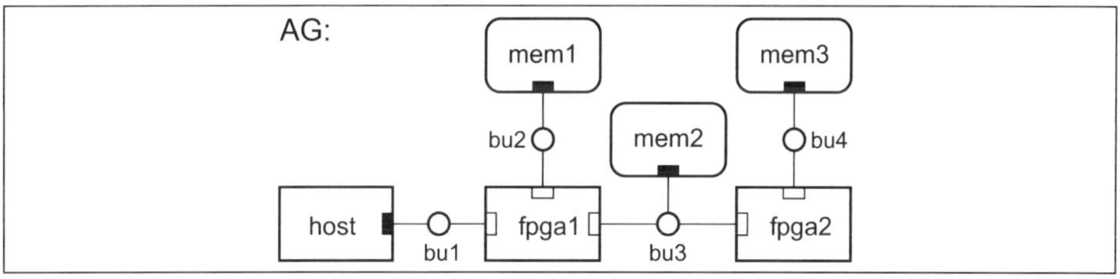

Figure 5.11: Architecture graph.

As described in Section 3.3.3, the mapping of the problem graph to the architecture graph takes four steps:

Step 1: *Define potential assignments to configurations and components* Assume that there exist six configurations $\Delta = \delta_1 \ldots \delta_6$. Now, the potential node-configuration assignments ϕ^* and potential configuration-component assignments ψ^* have to be specified (see Figure 5.12b). They depend on the potential binding β^*. For example, the buffer b_7 can be bound to $fpga2$. Therefore, buffer b_7 can be member of configurations δ_4 and δ_6 only.

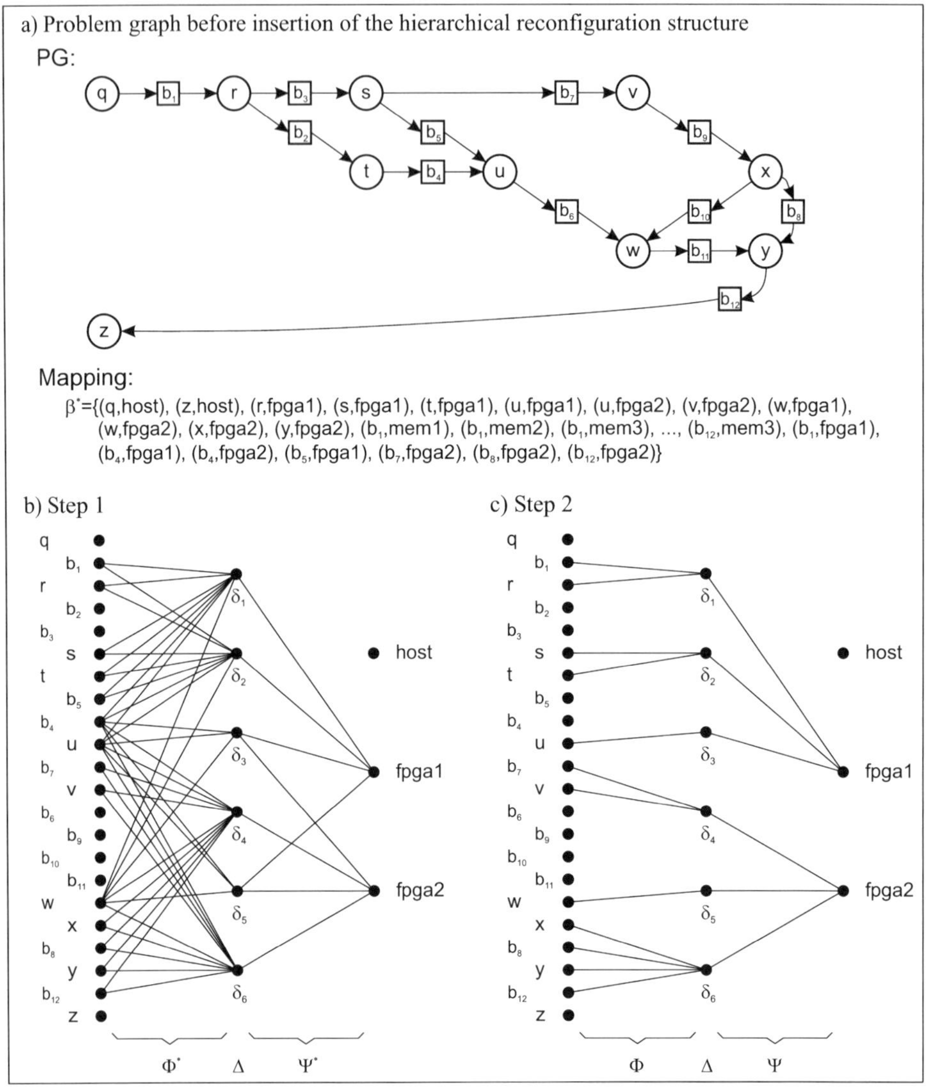

Figure 5.12: Problem specification.

Step 2: *Select appropriate assignments*

Based on step 1, a node-configuration assignment ϕ as well as a configuration-component assignment ψ has to be selected (see Figure 5.12c). Note, that this selection actually depends on the fire schedule of the problem graph nodes which has been considered here. The schedule is specified in Figure 5.10.

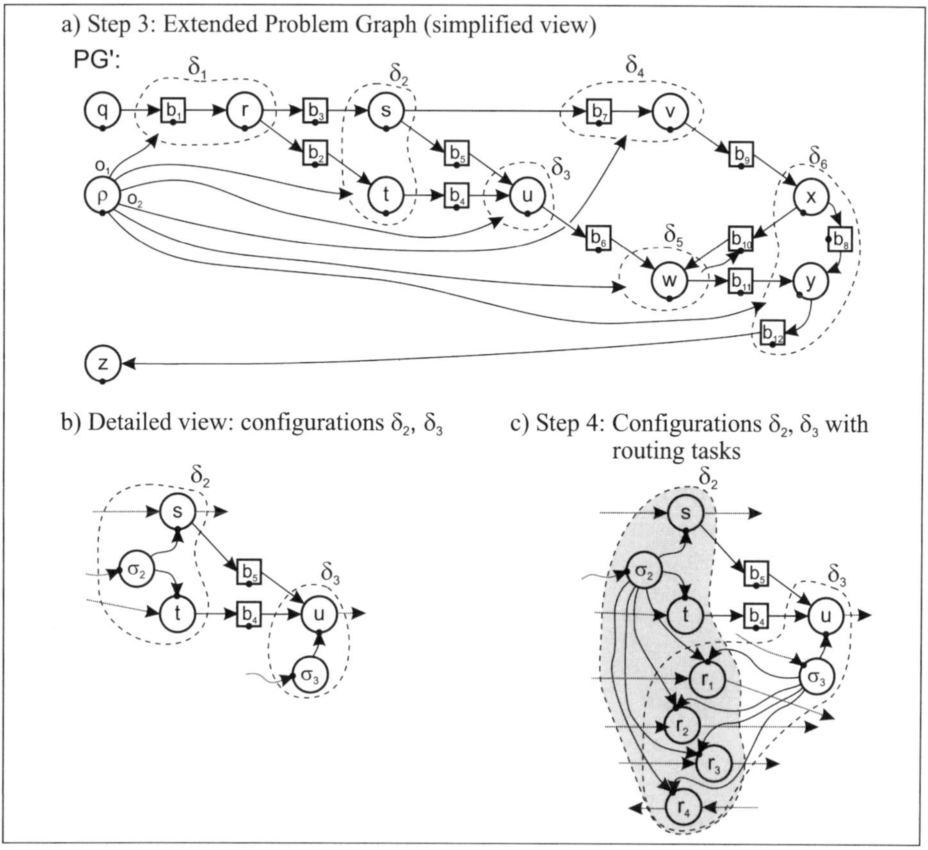

Figure 5.13: Problem specification: Extended problem graph.

Step 3: *Insert the hierarchical reconfiguration structure*

Here, an additional dispatcher node has to be inserted for each used configuration, as well as at least one configurator node bound to the host computing resource. Subsequently, an extended problem graph PG' arises (see Figure 5.13a). The configurator node ρ is connected to the control input of each inserted dispatcher node which is only indicated in Figure 5.13a). The dispatcher of a configuration is connected to the control inputs of all remaining nodes of the configuration. As an example, Figure 5.13b) shows a detailed view of configurations δ_2 and δ_3.

Step 4: *Consider constraints*

The constraints assure a correct specification. In this example, constraint *C3* (see Section 3.2.3) is violated for the edges between the configurator

ρ and the dispatchers of configurations $\delta_4 \ldots \delta_6$ as well as for the edge $b_{12} - z$. Each path for these edges has a length of 4. Subsequently, routing tasks $r_1 \ldots r_4$ have to be inserted onto these edges which finally satisfies constraint *C3*. As a consequence, these tasks on *fpga1* are shared between the configurations (see Section 3.3.5 for sharing of tasks). Figure 5.13c) shows again configurations δ_2 and δ_3 where the routing tasks have been inserted. For the sake of clarity, the shaded area denotes the nodes of configuration δ_2.

5.4.2 Problem Refinement

To establish a communication channel between a sender and a receiver the corresponding edge has to be refined (see Section 4.1.1). Subsequently, channel access points have to be inserted which results in a refined problem graph rPG. Figure 5.14 shows parts of this graph for configurations

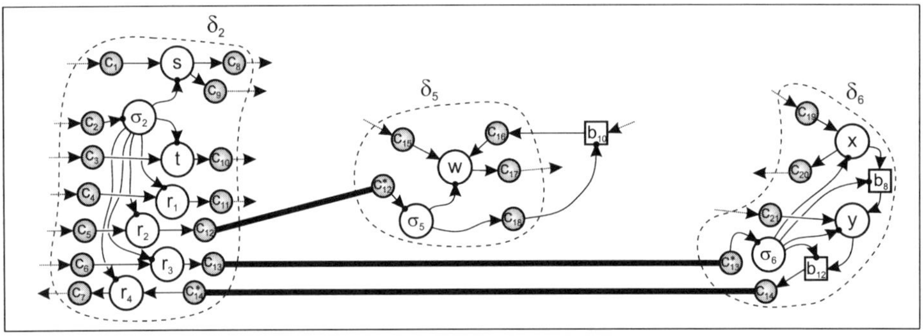

Figure 5.14: Parts of the refined problem graph rPG.

δ_2, δ_5, and δ_6, as well as for some intermediate edges. Bold lines denote connections between channel access points. In the original SDF graph (see Figure 5.10) one edge features an initial marking. Subsequently, the buffer b_{10} has to be initialized. This is modeled by a channel between the dispatcher σ_5 and the control input of buffer b_{12}.

5.4.3 Optimization and Synthesis

Optimization

Prior to the synthesis step, optimization methodologies are applied (see Section 5.2). Subsequently, the number of shared nodes and required pins to implement the communications channels are minimized. The optimization makes use of the fact that configurations $\delta_1 ... \delta_3$ and $\delta_4 ... \delta_6$ are time-exclusive.

The first optimization considers the channels between the configurator ρ and the dispatchers $\sigma_1 \ldots \sigma_3$ of configurations $\delta_1 ... \delta_3$. As these channels are used exclusively, *interconfiguration* pin sharing can be applied. Hence, actually only one channel has to be implemented.

The second optimization tackles the routing tasks required to connect the configurator ρ and the dispatchers $\sigma_4 \ldots \sigma_6$ of the configurations $\delta_4 \ldots \delta_6$. Again, it never happens that data have to be transmitted via these three nodes at the same time. Therefore, only one task is physically necessary that implements all three logically required routing tasks. Therefore, *intra-configuration* pin sharing can be applied for the channels and *node sharing* for the routing tasks. Again, only one channel has to be implemented.

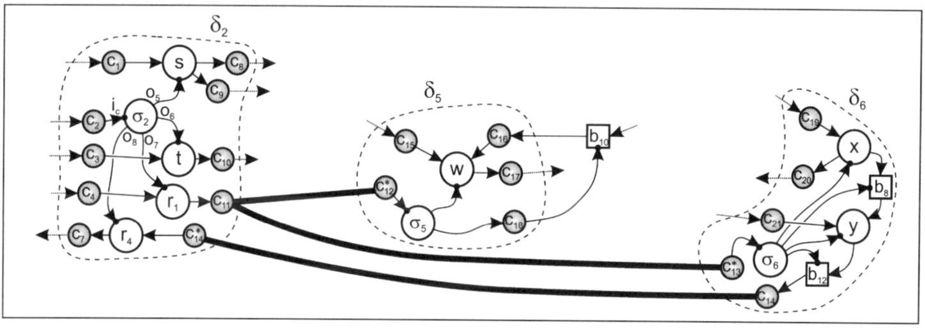

Figure 5.15: Optimized refined problem graph rPG'.

Figure 5.15 outlines the same part of the refined problem graph as shown in Figure 5.14 where the optimizations have been applied. Note that the channels access points c_{12}^* and c_{13}^* will have an instance of the same interface circuit. Therefore, switching between the configurations $\delta_4 \ldots \delta_6$ connects one of the dispatchers with the routing task r_1 and enables the supervision of the configuration by the configurator on the host. Switching between the configurations $\delta_1 \ldots \delta_3$ interrupts the channel between the configurator and a dispatcher on *fpga2* during reconfiguration time. It is reestablished as soon as the reconfiguration of *fpga1* is finished and the routing task is started.

The above optimizations reduce the number of necessary channels for the hierarchical reconfiguration structure. Subsequently, the configurator

ρ requires only two outputs o_1 and o_2 to supervise the dispatchers. The dispatchers $\sigma_1 \ldots \sigma_3$ require two outputs for the routing tasks.

Synthesis

During synthesis interface circuits and device drivers are generated and assembled according to the specification of the configurations. To implement the schedule of the original SDF, the configurator ρ as well as the dispatchers $\sigma_1 \ldots \sigma_6$ have to be implemented. Figure 5.16a) shows a pseudo code for the configurator FSM of the configurator ρ. It permanently reconfigures both FPGA resources upon end of configuration using it's output ports o_1 and o_2 (see Figure 5.13). Figure 5.16b) shows a pseudo code for the dispatcher FSM of configuration δ_2 using its output ports $o_5 \ldots o_8$ (see Figure 5.15).

a) Configurator FSM (ρ)

```
cFSM() {
  F1list={δ1,δ2,δ3}; F2list={δ4,δ5,δ6}
  F1cfg=F1list.first(); F2cfg=F2list.first()
  download1(F1cfg); o1.start()
  download2(F2cfg); o2.start()
  loop {
    case o1.done():
      if F1cfg==F1list.last()
        F1cfg=F1list.first()
      else
        F1cfg.next()
      download1(F1cfg); o1.start()
    case o2.done():
      if F2cfg==F2list.last()
        F2cfg=F2list.first()
      else
        F2cfg.next()
      download2(F2cfg); o2.start()
  }
}
```

b) Dispatcher FSM (σ_2)

```
dFSM() {
  if ic.start() {
    o5.start(); o6.start()
    o7.start(); o8.start()
    loop {
      case o5.done() & o6.done()
        o7.stop(); o8.stop()
        if (o7.done() & o8.done())
          exit loop
      case o7.done():
        o7.start()
      case o8.done():
        o8.start()}
    ic.done()
  }
}
```

Figure 5.16: Pseudo codes for a) configurator ρ, and b) dispatcher σ_2.

Run-time overheads

The hierarchical reconfiguration structure as well as the channel access points introduce overheads in terms of *execution time* and *hardware area*. Both features strongly depend on the target technology as well as compiler methodologies for the computing resources.

Generally, the *execution time* depends on (i) the execution order of configurations, (ii) the run-time of each configuration, and (iii) the switching time between two configurations. The configuration's execution order is captured by the configurator FSM and depends on the schedule of problem graph nodes. The configuration's run-time is determined by (i) the run-time of nodes and (ii) by the execution order of problem graph nodes assigned to this configuration. This execution order is captured by the dispatcher of the configuration. The switching time t_{xy} from a configuration δ_x to a configuration δ_y represents *run-time overhead* and extends the run-time compared to CTR resources. It is the sum of t_e, the time required by the dispatcher to wait for the end of configuration after stopping the connected nodes, t_t, the transmission delay between the dispatcher and the corresponding configurator, t_s, the scheduling time of the cFSM, and t_{conf}, the time for downloading and starting the new configuration.

$$t_{xy} = t_e + t_t + t_s + t_{conf}.$$

For specification models with rather simple schedules, t_e and t_s will be small. Assuming moderately sized multi-FPGA systems, t_t will be small as communication channels between dispatchers and configurators are not routed over many FPGAs. For such a scenario, the configuration switching time will be dominated by the technology-dependent device reconfiguration time t_{conf}. For example, an XILINX Virtex XCV-1000 FPGA requires about 6.13 Mbits of configuration data. By reconfiguring the device at 50 MHz using the fastest programming port (8 bit) the reconfiguration time is about $t_{conf} = 15.3ms$.

The *area overhead* required for the communication infrastructure consists of (i) the generated interfaces and device drivers to implement the communication channnels, (ii) the configurator and dispatcher tasks for the hierarchical reconfiguration, (iii) the required routing tasks, and (iv) the wire resources to connect the implementations. Besides the bit width of a channel implementation, the area requirements of interfaces and device drivers depend on the computing resource type as well as the selected protocol of an edge. The size of configurators and dispatcher tasks depends on the number of states and transitions required to implement a schedule. Routing tasks as well as wire resources depend mainly on the bit width of a channel.

Measurements on XILINX Virtex XCV-1000 FPGAs (programmable matrix provides 6144 CLBs) show that the area overhead due to *dynamic reconfiguration* for the discussed example is quite small. Tab. 5.1 summarizes some of the area overheads expressed in Virtex CLBs (configurable logic blocks). Here, the channel's implementation has a data width of 8 bit.

additional element	Virtex CLBs
routing task (e.g., r_4)	5
dispatcher (e.g., σ_2)	8
interface between *fpga1*-*fpga2*	2
interface between *host*-*fpga1*	5

Table 5.1: Area overhead due to dynamic reconfiguration.

For example, for configuration δ_2 in Figure 5.15 the area overhead due to dynamic reconfiguration is $2 * 5 + 8 + 2 * 2 + 3 * 5 = 37$ CLBs which is 0.6% of the available CLBs of a XILINX Virtex XCV-1000 FPGA.

5.5 Summary

This chapter summarizes the proposed models and refinements in a coherent view.

It presents a synthesis flow that consists of three phases. At first, a problem is specified using the EPS or the RPS formalism. Next, by the insertion of channel access points the specification is refined and now captures the specification of the communication infrastructure as well. The last phase optimizes and synthesizes the specification to source code for each computing resource and configuration.

The next section presents an optimization technique to reduce the number of problem graph nodes and required device pins to implement the communication channels. Essentially, it bases on the fact some of the configurations are exclusive.

Furthermore, a framework is outlined that has been developed during the research work. It implements the proposed synthesis flow and bases on the object-oriented component model.

Finally, an extended example is elaborated that presents specification, refinement, and synthesis of a simple SDF graph.

Chapter 6

Conclusions

6.1 Results

The goal of this thesis was the definition of models and methodologies to establish a communication infrastructure within heterogeneous embedded systems. Such systems are characterized by connected devices that distinguish in features such as supported interfaces and the ability of run-time reconfiguration. The essential results can be summarized as follows:

- To capture and describe synthesis problems three dataflow oriented specification models have been formalized. Their focus is on the specification of point-to-point communication channels between interacting parts. The GPS formalism is the parent model and provides definition of terms, basic elements, and composition rules. The EPS formalism is an extension of GPS and aims at static configured systems. The RPS formalism is an extension of EPS and supports the specification of systems with run-time reconfigurable components.

- For synthesis problems specified by using the EPS or the RPS formalism refinement and implementations methodologies have been proposed. They consider the communication features of architecture components by the introduction of channel access points into the behavior specification.

- A taxonomy of communication types has been proposed that is based solely on (i) the binding of problem graph nodes to basic architecture components, and (ii) on the existence of one or several configurations.

- To establish a communication infrastructure for heterogeneous embedded systems object-oriented models for the components of the architecture graph have been proposed. Essentially, the communication features of a device are captured by a set of classes that enable the generation of interface circuitry and device drivers. This object-oriented approach is the base for (i) simple component modeling, (ii) reuse of existing generation methodologies, (iii) simple retargeting, and (iv) object interaction.

- The specification as well as the synthesis process of problem graph nodes is independent of a hardware or software implementation of the nodes.

- To reduce the overhead (FPGA area, and device pins) caused by the dynamic reconfiguration an optimization methodology has been proposed. It makes use of the fact that some configurations are exclusive.

6.2 Future Perspectives

Based on the gained results, further research may include:

- *Extension of the chip models*

  Currently, the chip models of the object-oriented approach model the communication features of the corresponding component. It could be useful to extend these models by further device features such as on-chip timers, and CPU core. Such an extended model provides a more realistic view of the devices and could be used for a refined synthesis process.

- *Communication scheduling*

  During the refinement of the problem graph channel access points are inserted. These nodes should be included into an overall system schedule. Furthermore, the additional configurator, dispatcher, and routing nodes have to be included as well.

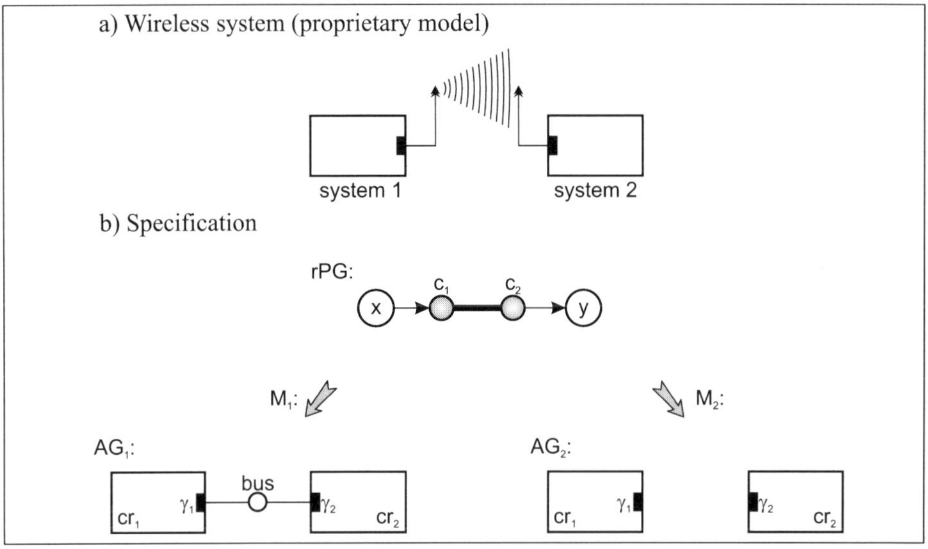

Figure 6.1: Wireless system.

- *Broken channels*

  The specification models base on a channel model where the channel connects a sender and a receiver by using a blocking or non-blocking access semantic. This model assumes that sent data actually arrive at a receiver. However, this model is not sufficient for channels that may break, i.e., where sender or receiver may not reachable. Consider a simple wireless application (see Fig. 6.1a) where the two systems communicate via a radio transmission. A first specification approach is given in Fig. 6.1b) where a problem graph is fixed but the architecture graph as well as the mapping changes dynamically. Here, the formalism requires a communication semantics where an issued sending/receiving of data items can be withdrawn to avoid deadlock situations. A step in this direction could be an access semantic where sending/receiving is on a trial basis.

Appendix A

Paper Summary

The research results of this thesis have been published at several workshops, conferences, and journals. In the following, a short overview is given:

HICSS 98: [ETT98] tackles the problem of automatically mapping large-grain dataflow programs onto heterogeneous hardware/software architectures. The paper introduces simplified problem and architecture models, outlines the implementation of hardware/software channels using interface circuitry and device drivers, and presents the wrapping of IP (Intellectual Property) cores.

CODES 98: [ET98a] introduces the HASIS (**H**ardware **S**oftware **In**terface **S**ynthesis) tool that bases on an object-oriented component model for the devices of a target architecture. Using such components, a simple methodology of communication channel synthesis is proposed.

FPL 98: [ET98b] treats the automatic generation of communication channels for heterogeneous embedded systems. It describes an efficient synchronization protocol that is used for generic interface/device driver templates. Additionally, techniques for reducing the power consumption for nodes on hardware computing resources are outlined.

FPL 99: [EPT99] presents a methodology for the communication synthesis of reconfigurable systems based on problem and architecture graphs. A taxonomy for compile- and run-time reconfigurable systems outlines possible communication types. Furthermore, issues of communication between exclusive FPGA configurations are discussed.

IEE Proceedings - Computers and Digital Techniques: [EP00c] elaborates the results of [EPT99] in more detail. Furthermore, it presents CORES (**Co**nfigurator for **R**econfigurable **E**mbedded **S**ystems), a set of tools that enables the synthesis of interface circuitry for run-time reconfigurable systems. Additionally, the problem of reconfiguring non-adjacent computing resources is discussed.

IEEE Design and Test of Computers: [EZT00] presents the design and implementation of complex embedded systems using a hybrid evolutionary algorithm. Here, the focus is on finding optimal target architectures for a given problem graph considering several competing objectives such as latency, communication overhead, etc.

ENREGLE 2000: [EP00a] extends the framework in [EP00c] and elaborates the main issues of reconfiguration synthesis. This includes the introduction of a formalism for the specification of reconfigurable systems as well as reconfiguration control.

FPL 2000: [EP00b] presents optimization techniques for reconfigurable systems based on the results published in [EP00a]. The optimizations target the reconfiguration structure and its communication requirements and base on the sharing of objects such as device pins and problem graph nodes.

Kluwer Journal of Supercomputing: [EP02] outlines the design and implementation of run-time reconfigurable systems. Essentially, it describes the RPS problem specification, the object-oriented component model, as well as the hierarchical reconfiguration structure.

Curriculum Vitae

Name	Michael Herbert Eisenring
Date of birth	July 31, 1967 in Wettingen, Switzerland

Education

1974 - 1983	Elementary and secondary school, Dübendorf, Switzerland
1983 - 1987	Serve an apprenticeship as FEAM (Fernmelde- und Elektronikapparatemonteur) at Zellweger Uster AG, Uster, Switzerland
1987 - 1990	Studies of electrical engineering at the University of Applied Sciences Rapperswil (ITR), Switzerland, Degree as Electrical Engineer HTL
1990 - 1991	Business management education at IFKS Zurich, (Institut für Kaderschulung), Business management certificate IFKS
1992 - 1996	Studies of electrical engineering at ETH Zurich, Degree as Electrical Engineer ETH
1996 - 2002	Research assistant and Ph.D. student at the computer engineering and networks laboratory, department of electrical engineering, ETH Zurich

Professional Experience

April-Nov., 1987	FEAM at Zellweger Uster AG
1990 - 1992	Employment as hardware/software engineer for embedded systems at SYSTAG AG, Rüschlikon, Switzerland
since June, 2000	CEO of B2B Consulting AG, Swiss representative of Fairchild Semiconductor Inc.
since Nov., 2000	Lectureship for software (Algorithms and Data Structures) at University of Applied Sciences Winterthur (ZHWIN), Computer Science Department

Bibliography

[AGH00] Ken Arnold, James Gosling, and David Holmes. *The JAVA programming language*. Addison-Wesley, 2000.

[AH95] S. AlKasabi and S. Hariri. A dynamically reconfigurable switch for high-speed networks. In *IEEE 14th Annual International Phoenix Conference on Computers and Communications*, pages 508–514, March 28–31 1995.

[ALT] ALTERA. http://www.altera.com.

[Apt] Aptix. Aptix: Reconfigurable System Prototyping. http://www.aptix.com.

[ARM] ARM. AMBA: Advanced Microcontroller Bus Architecture. http://www.arm.com/sitearchitek/armtech.ns4/html/amba.

[BA90] M. Ben-Ari. *Principles of Concurrent and Distributed Programming*. Prentice Hall, 1990.

[BCG+97] F. Balarin, M. Chiodo, P. Giusto, H. Hsieh, A. Jurecska, L. Lavagno, C. Passerone, A. Sangiovanni-Vincentelli, E. Sentovich, K. Suzuki, and B. Tabbara. *Hardware-Software Co-Design of Embedded Systems: The Polis Approach*. Kluwer Academic Press, June 1997.

[BCO96] G. Borriello, P. Chou, and R. Ortega. *Embedded System Co-Design: Towards Portability and Rapid Integration*. Hardware/Software Co-Design, Kluwer Academic Publishers, pages 234–264, 1996.

[BEK+95] T. Benner, R. Ernst, I. Könenkamp, P. Schüler, and H. Schaub. A Prototyping System for Verification and Evaluation in Hardware-Software Cosynthesis. In *6th International Workshop on Rapid System Prototyping*, pages 54–59, Chapel Hill, North Carolina, USA, June 1995.

[BG00] J. Becker and M. Glesner. Ip-based application mapping techniques for dynamically reconfigurable hardware architectures. In *Second International Workshop on Engineering of Reconfigurable Hardware/Software Objects (ENREGLE)*, pages 221–227, Las Vegas, Nevada, USA, June 26–29 2000.

[BH98] P. Bellows and B. Hutchings. JHDL – An HDL for Reconfigurable Systems. In *IEEE Symposium on FPGAs for Custom Computing Machines*, pages 175–184, Los Alamitos, CA, April 1998. IEEE Computer Society Press, http://www.jhdl.com.

[BHLM91] J. Buck, S. Ha, E. Lee, and D. Messerschmitt. Ptolemy: A framework for simulating and prototyping heterogeneous systems. *International Journal on Computer Simulation*, 4:155–182, 1991.

[Bor88] Gaetano Borriello. *A New Interface Specification Methodology and its Application to Transducer Synthesis*. PhD thesis, Computer Science Division (EECS), University of California Berkely, California 94720, 1988. Report No. UCB/CSD 88/430.

[Bor92] G. Borriello. Formalized Timing Diagrams. In *European Conference on Design Automation*, pages 372–377, Brussels, Belgium, March 16–19 1992.

[BP95] D. Buell and K. Pocek. Custom Computing Machines: An Introduction. *Journal of Supercomputing*, 9(3):219–229, 1995.

[BR95] K. Buchenrieder and J. Rozenblit. *Codesign: Computer-Aided Software/Hardware Engineering*. IEEE Press, 1995.

[BTA93] J. Babb, R. Tessier, and A. Agarwal. Virtual wires: Overcoming pin limitations in fpga-based logic emulators. In *IEEE Workshop on FPGA-based Custom Computing Machines*, pages 142–151, Napa, CA, April 1993.

[BTD+97] J. Babb, R. Tessier, M. Dahl, S. Hanono, D. Hoki, and A. Agarwal. Logic emulation with virtual wires. *IEEE Transactions on Computer-Aided Design of Integrated Circuit and Systems*, 16(6):609–626, June 1997.

[BTT98] T. Blickle, J. Teich, and L. Thiele. System-level synthesis using evolutionary algorithms. *Design Automation for Embedded Systems, Kluwer Academic Publishers*, 3(8):23–58, January 1998.

[Buc93] J. T. Buck. Scheduling dynamic dataflow graphs with bounded memory using the Token Flow Model. Technical Report UCB/ERL 93/69, Ph.D dissertation, Department of EECS, UC Berkeley, Berkeley, CA 94720, U.S.A., 1993.

[BWE+93] G. Bilsen, P. Wauters, M. Engels, R. Lauwereins, and J. Peperstraete. Development of a static load balancing tool. In *Proceedings of the fourth Workshop on Parallel and Distributed Processing*, pages 179–194, Sofia, Bulgaria, 1993.

[CH71] F. Commoner and A. Holt. Marked directed graphs. *Journal of Computer and System Sciences*, 5:511–523, 1971.

[CH00] K. Compton and S. Hauck. Reconfigurable Computing: A Survey of Systems and Software. *Submitted to ACM Computing Surveys*, 2000.

[Cie] Cierto. Cierto Virtual Component Co-Design. CADENCE, http://www.cadence.com/technology/hwsw/ciertovcc/.

[COB92] Pai Chou, R. Ortega, and G. Borriello. Synthesis of the Hardware/Software Interface in Microcontroller-Based Systems. In *IEEE/ACM International Conference on Computer-Aided Design,*, pages 488–495, November 1992.

[COB95a] P. Chou, R. Ortega, and G. Borriello. Interface Co-Synthesis Techniques for Embedded Systems. In *IEEE/ACM International Conference on Computer-Aided Design*, pages 280–287, San Jose, CA, November 1995.

[COB95b] P. Chou, R. Ortega, and G. Borriello. The Chinook Hardware/Software Co-Synthesis System. In *8th International Symposium on System Synthesis*, pages 22–27, September 13–15 1995.

[COH+99] P. Chou, R. Ortega, K. Hines, K. Partridge, and G. Borriello. ipChinook: An Integrated IP-based Design Framework for Distributed Embedded Systems. In *36th Design Automation Conference*, pages 44–49, New Orleans, LA, June 21–25 1999.

[COS] COSSAP. COSSAP System Level Design. SYNOPSYS, http://www.synopsys.com/products/dsp/dsp.html.

[CV99] K. Chatha and R. Vemuri. Hardware-Software Codesign for Dynamically Reconfigurable Architectures. In *9th International Workshop on Field-Programmable Logic and Applications, FPL'99, Lecture*

Notes in Computer Science, 1673, pages 175–184, Glasgow, UK, August/September 1999.

[DIJ95] J. Daveau, T. Ismail, and A. Jerraya. Synthesis of System-Level Communication by an Allocation-Based Approach. In *8th International Symposium on System Synthesis*, pages 150–155, September 13–15 1995.

[DJ98] R. Dick and N. Jha. CORDS: Hardware-Software Co-Synthesis of Reconfigurable Real-Time Distributed Embedded Systems. In *IEEE/ACM International Conference on Computer-Aided Design*, pages 62–68, November 8–12 1998.

[DMBIJ97] J. Daveau, G. Marchioro, T. Ben-Ismail, and A. Jerraya. Protocol Selection and Interface Generation for HW-SW Codesign. *IEEE Transactions on Very Large Scale Integration (VLSI) Systems*, 5(1):136–144, March 1997.

[DZ83] J. Day and H. Zimmermann. The OSI Reference Model. *Proceedings of the IEEE*, 71(12):1334–1340, 1983.

[EBLP94] M. Engels, G. Bilsen, R. Lauwereins, and J. Peperstraete. Cyclo-Static Data Flow: Model and implementation. In *Proceedings 28th Asilomar Conference on Signals, Systems, and Computers*, pages 503–507, Pacific Grove, CA, 1994.

[EHB+96] R. Ernst, J. Henkel, T. Benner, W. Ye, U. Holtmann, and D. Hermann. The COSYMA environment for hardware/software cosynthesis of small embedded systems. *Microprocessors and Microsystems*, 20(3):159–166, May 1996.

[Eis96] M. Eisenring. Hardware/Software Codesign in a Microcontroller and FPGA based System. Masters thesis 1996, ETH Zurich, Computer Engineering Lab, 1996.

[Eis99] M. Eisenring. CCSL, Communication Channel Synthesis Language. TIK Report No. 80, Computer Engineering and Networks Laboratory, ETH Zurich, Switzerland, 1999.

[ELSS94] K. Etschberger, A. Lorinser, C. Schlegel, and T. Suters. *CAN, Controller Ara Network*. Carl Hanser Verlag, 1994.

[Eon99] Eonic. Virtuoso, Real-Time Software Development Tools for Embedded Systems. http://www.eonic.com/, 1999.

[EP00a] M. Eisenring and M. Platzner. An Implementation Framework for Run-Time Reconfigurable Systems. In *The Second International Workshop on Engineering of Reconfigurable Hardware/Software Objects, ENREGLE 2000*, pages 151–157, Las Vegas, Nevada, USA, June 26–29 2000.

[EP00b] M. Eisenring and M. Platzner. Optimization of Run-Time Reconfigurable Embedded Systems. In *10th International Workshop on Field Programmable Logic and Applications, FPL 2000*, pages 565–574, Villach, Austria, August 28–30 2000.

[EP00c] M. Eisenring and M. Platzner. Synthesis of Interfaces and Communication in Reconfigurable Embedded Systems. *IEE Proceedings – Computers and Digital Techniques*, 147(3):159–165, May 2000.

[EP02] M. Eisenring and M. Platzner. A Framework for Run-time Reconfigurable Systems. *The Journal of Supercomputing, Kluwer Academic Publishers*, 21(2):145–159, February 2002.

[EPT99] M. Eisenring, M. Platzner, and L. Thiele. Communication Synthesis for Reconfigurable Embedded Systems. In *9th International Workshop on Field-Programmable Logic and Applications*, pages 205–214, Glasgow, UK, August/September 1999.

[Ern97a] R. Ernst. Hardware/Software Co-Design of Embedded Systems. In *Asia Pacific Conference on Computer Hardware Description Languages (APCHDL)*, Hsin-Chu, Taiwan, Aug. 18–20 1997.

[Ern97b] R. Ernst. Hardware/software co-design of signal processing systems. In *IEEE workshop on Signal Processing Systems (SIPS)*, Leicester, UK, Nov. 3–5 1997.

[Ern98] R. Ernst. Embedded System Architectures. In *System-Level Synthesis*, volume 357 of *Applied Sciences*, pages 1–43. Il Ciocco, Nato Advanced Study Institute on System Synthesis, August 1998.

[ET98a] M. Eisenring and J. Teich. Domain-Specific Interface Generation from Dataflow Specifications. In *Sixth International Workshop on Hardware/Software Codesign*, pages 43–47, Seattle, WA, March 1998.

[ET98b] M. Eisenring and J. Teich. Interfacing Hardware and Software. In *8th International Workshop on Field-Programmable Logic and Applications, FPL'98, Lecture Notes in Computer Science, 1482*, pages 520–524, Tallinn, Estonia, August 31–September 3 1998.

[ETT98] M. Eisenring, J. Teich, and L. Thiele. Rapid Prototyping of Dataflow Programs on Hardware/Software Architectures. In *Proceedings of HICSS-31, Proceedings of the Hawaii International Conference on Syst. Sci.*, volume VII, pages 187–196, Kona, Hawaii, January 1998.

[Exc] Excalibur Backgrounder, White Paper. http://www.altera.com/html/products/excalibursplash.html.

[EZT99] M. Eisenring, E. Zitzler, and L. Thiele. CoFrame: A Modular Co-Design Framework for Heterogeneous Distributed Systems. TIK Report No. 81, Computer Engineering and Networks Laboratory, ETH Zurich, Switzerland, 1999.

[EZT00] M. Eisenring, E. Zitzler, and L. Thiele. Conflicting Criteria in Embedded System Design. *IEEE Design and Test of Computers*, 17(2):51–59, April–June 2000.

[FGSS98] W. Fornaciari, P. Gubian, D. Sciuto, and C. Silvano. Power Estimation of Embedded Systems: A Hardware/Software Codesign Approach. *IEEE Transactions on Very Large Scale Integration (VLSI) Systems*, 6(2):266–275, June 1998.

[FSS99] W. Fornaciari, D. Sciuto, and C. Silvano. Power Estimation for Architectural Exploration of HW/SW Communication on System-Level Buses. In *Seventh International Workshop on Hardware/Software Codesign*, pages 152–156, Rome, Italy, May 3–5 1999.

[FSV99] A. Ferrari and A. Sangiovanni-Vincentelli. System Design: Traditional Concepts and New Paradigms. In *IEEE International Conference on Computer Design*, pages 2–12, Austin, Texas, October 10–13 1999.

[GDZ99] D. Gajski, R. Dömer, and J. Zhu. IP-Centric Methodology and Specification Language. *Distributed and Parallel Embedded Systems, Edited by Franz J. Rammig, Kluwer Academic Press, Boston*, pages 3–21, 1999.

[GG95] S. Guccione and M. Gonzalez. Classification and Performance of Reconfigurable Architectures. In *Field-Programmable Logic and Applications*, pages 439–448. Springer-Verlag, Berlin, August/September 1995. 5th International Workshop on Field-Programmable Logic and Applications, FPL 1995. Lecture Notes in Computer Science 975.

[GHJV95] E. Gamma, R. Helm, R. Johnson, and J. Vlissides. *Design Patterns.* Addison-Wesley, 1995.

[GL99] S. Guccione and D. Levi. Design Advantages of Run-Time Reconfiguration. In *Reconfigurable Technology: FPGAs for Computing and Applications, Proceedings SPIE 3844*, pages 87–92, Bellingham, WA, September 1999. SPIE – The International Society for Optical Engineering.

[GM93] R. Gupta and G. De Micheli. Hardware-Software Co-Synthesis of Digital Systems. *IEEE Design and Test of Computers*, 10(3):29–41, September 1993.

[Gon97] J. Gong. Model Refinement for Hardware-Software Codesign. *ACM Transactions on Design Automation of Electronic Systems*, 2(1):22–41, January 1997.

[Gra] Mentor Graphics. http://www.mentor.com.

[GV00] T. Givargis and F. Vahid. Parameterized System Design. In *Eighth International Workshop on Hardware/Software Codesign*, pages 98–102, San Diego, California, May 3–5 2000.

[GVNG94] D. Gajski, F. Vahid, S. Narayan, and J. Gong. *Specification and Design of Embedded Systems.* Prentice Hall, 1994.

[Har87] D. Harel. Statecharts: A visual formalism for complex systems. *Science of Computer Programming*, 8:231–274, 1987.

[HB97] S. Hauck and G. Borriello. Pin Assignment for Multi-FPGA Systems. *IEEE Transactions on Computer-Aided Design of Integrated Circuit and Systems*, 16(9):956–964, September 1997.

[HBK96] R. Hartenstein, J. Becker, and R. Kress. Custom computing machines versus hardware/software co-design: from a globalized point of view. In *Proceedings 6th International Workshop on Field Programmable Logic and Applications, FPL'96. Lecture Notes in Computer Science, Springer Press*, Darmstadt, Germany, September 1996.

[HBKG98] T. Hollstein, J. Becker, A. Kirschbaum, and M. Glesner. HiPART: A New Hierarchical Semi-Interactive HW/SW Partitioning Approach with Fast Debugging for Real-Time Embedded Systems. In *Sixth International Workshop on Hardware/Software Codesign*, pages 29–33, Seattle, WA, March 1998.

[HBS98] J. Haenni, J. Beuchat, and E. Sanchez. RENCO: A Reconfigurable Network Computer. In *IEEE Symposium on FPGAs for Custom Computing Machines*, pages 288–289, Napa Valley, California, April 15–17 1998.

[HFHK97] S. Hauck, T. Fry, M. Hosler, and J. Kao. The Chimaera Reconfigurable Functional Unit. In *IEEE Symposium on FPGAs for Custom Computing Machines*, pages 87–96, Napa Valley, California, April 1997.

[HLS98] S. Hauck, Z. Li, and E. Schwabe. Configuration Compression for the XILINX XC6200 FPGA. In *IEEE Symposium on FPGAs for Custom Computing Machines*, pages 138–146, Napa Valley, California, April 15–17 1998.

[HSS00] Marco Hauri, Rolf Sigg, and Thomas Singer. Communication and reconfiguration on heterogeneous computing platforms. Diploma thesis, ETH Zurich, Computer Engineering and Networks Laboratory, February 2000. Supervisor: M. Eisenring.

[HW95] B. Hutchings and M. Wirthlin. Implementation Approaches for Reconfigurable Logic Applications. In *International Workshop on Field-Programmable Logic and Applications*, pages 419–428, 1995.

[Idt] Integrated Device Technology Idt. http://www.idt.com.

[IEE87] IEEE. *IEEE Std 1014–1987, IEEE Standard for a Versatile Backplane Bus: VMEbus*. IEEE, 1987.

[Inc] Chameleon Systems Inc. Reconfigurable Communications Platform. http://www.chameleonsystems.com/.

[Ins] Texas Instruments. Texas Instruments. http://www.ti.com.

[Int] Intel. http://www.intel.com.

[IS95] C. Iseli and E. Sanchez. Spyder: A SURE (SUperscalar and REconfigurable) processor. *Journal of Supercomputing*, 9(3):231–252, 1995.

[Jan00] J. Janneck. *Syntax and Semantics of Graphs – An approach to the specification of visual notations for discrete event systems*. Phd thesis, ETH Zurich, Computer Engineering and Networks Laboratory, July 2000.

[JO95] A. Jerraya and K. O'Brien. *Solar: An Intermediate Format for System-Level Modeling and Synthesis*, chapter 7, pages 145–175. IEEE Press, Codesign: Computer-Aided Software/Hardware Engineering edition, 1995. K. Buchenrieder and J. Rozenblit (eds).

[Kah74] G. Kahn. The semantics of a simple language for parallel programming. In *Proceedings of the IFIP Congress 74*, pages 471–475, North Holland, 1974.

[Kau00] T. Kaufmann. Application on a Multi-DSP/Multi-FPGA Platform. Diploma thesis, Computer Engineering and Networks Laboratory, 2000. Supervisor: M. Eisenring.

[Kea00] T. Kean. It's FPL, Jim–But not as We Know It! Opportunities for the New Commercial Architectures. In *10th International Workshop on Field Programmable Logic and Applications, FPL 2000*, pages 575–584, Villach, Austria, August 28–30 2000.

[KG97] A. Kirschbaum and M. Glesner. Rapid Prototyping of Communication Architectures. In *8th IEEE International Workshop on Rapid System Prototyping*, pages 136–141, Chapel Hill, North Carolina, USA, June 24–26 1997.

[KKR94] G. Koch, U. Kebschul, and W. Rosenstiel. A prototyping environment for hardware/software codesign in the COBRA project. In *Proceedings of Codes/CASHE'94 – the 3rd International Workshop on Hardware/Software Codesign*, pages 10–16, Grenoble, France, September 1994.

[KM98] P. Knudsen and J. Madsen. Communication Estimation for Hardware/Software Codesign. In *Proceedings of the Sixth International Workshop on Hardware/Software Codesign, CODES/CASHE'98, Seattle, Washington*, pages 55–59, March 15–18 1998.

[KV98] M. Kaul and R. Vemuri. Optimal temporal partitioning and synthesis for reconfigurable architectures. In *Proceedings of Design, Automation and Test in Europe*, pages 389–396, February 23–26 1998.

[LCD+00] Y. Li, T. Callahan, E. Darnell, R. Harr, U. Kurkure, and J. Stockwood. Hardware-Software Co-Design of Embedded Reconfigurable Architectures. In *37th Design Automation Conference*, pages 507–512, LA, USA, June 5–9 2000.

[LM87a] E. Lee and D. Messerschmitt. Static scheduling of synchronous dataflow programs for digital signal processing. *IEEE Transactions on Computers*, C-36(1):24–35, 1987.

[LM87b] E. Lee and D. Messerschmitt. Synchronous dataflow. *Proceedings of the IEEE*, 75(9):1235–1245, 1987.

[LM95] E. Lemoine and D. Merceron. Run Time Reconfiguration of FPGA for Scanning Genomic DataBases. In *IEEE Symposium on FPGAs for Custom Computing Machines*, pages 90–98, 1995.

[LNTT01] J. Lockwood, N. Naufel, J. Turner, and D. Taylor. Reprogrammable Network Packet Processing on the Field Programmable Port Extender (FPX). In *ACM International Symposium on Field Programmable Gate Arrays*, pages 87–93, Monterey, CA USA, February 11–13 2001.

[LRV+96] B. Lin, K. Rompaey, S. Vercauteren, D. Verkest, I. Bolsens, and H. De Man. Designing Single Chip Systems. In *2nd International Conference on ASIC*, pages 6–11, October 1996.

[LSS99] S. Ludwig, R. Slous, and S. Singh. Implementing PhotoShop Filters in Vertex. In *9th International Workshop on Field-Programmable Logic and Applications*, pages 231–242, Glasgow, UK, August/September 1999.

[LV94] B. Lin and S. Vercauteren. Synthesis of concurrent system interface modules with automatic protocol conversion generation. In *IEEE/ACM international conference on Computer-aided design*, pages 101–108, November 6–10 1994.

[Mar99] P. Marchal. Field-programmable gate arrays. *Communications of the ACM*, 42(4):57–59, 1999.

[MBL+96] H. De Man, I. Bolsens, B. Lin, K. Van Rompaey, S. Vercauteren, and D. Verkest. *Hardware/Software Co-design: Co-Design of DSP Systems, pages 75–104*, volume 310. Kluwer Academic Publishers, NATO Advanced Study Institute (ASI) Series E, 1996.

[Mic] Sun MicroSystems. JavaBeans. http://java.sun.com/products/javabeans/.

[Mic96] G. De Micheli. *Hardware/Software Co-Design: Application Domains and Design Technologies.* Hardware/Software Co-Design, Kluwer Academic Publishers, pages 1–28, 1996.

[MMF98] O. Mencer, M. Morf, and M. Flynn. PAM-Blox: High performance FPGA design for adaptive computing. In *IEEE Symposium on FPGAs for Custom Computing Machines*, pages 167–174, Los Alamitos, CA, April 1998. IEEE Computer Society Press.

[Mot] Motorola. Motorola Semiconductor Products. http://www.mot-sps.com/products/index.html.

[Mot94] Motorola. *MC68340, Integrated processor with DMA, user's manual*, 1994.

[NTE99] Martin Naedele, L. Thiele, and M. Eisenring. Characterizing Variable Task Releases and Processor Capacities. In *Proceedings 14th IFAC World Congress, Beijing, China*, July 5–9 1999.

[OB98] R. Ortega and G. Borriello. Communication Synthesis for Distributed Embedded Systems. In *IEEE/ACM International Conference on Computer-Aided Design*, pages 437–444, San Jose, CA, November 1998.

[OJ97] M. O'Nils and A. Jantsch. Communication in Hardware/Software Embedded Systems – A Taxonomy and Problem Formulation. In *15th NORCHIP Seminar, Copenhagen, Denmark*, pages 67–74, November 1997.

[OLB98] R. Ortega, L. Lavagno, and G. Borriello. Models and Methods for HW/SW Intellectual Property Interfacing. In *NATO Advanced Study Institute on System-level Synthesis*, 1998.

[Pac97] P. Pacheco. *Parallel Programming with MPI.* Morgan Kaufmann Publishers, 1997.

[Pat00] Cameron Patterson. High Performance DES Encryption in Virtex FPGAs using JBits. In *To appear in IEEE Symposium on Field-Programmable Custom Computing Machines*, April 2000.

[PB98] K. Purna and D. Bhatia. Emulating large designs on Small Reconfigurable Hardware. In *9th IEEE International Workshop on Rapid System Prototyping*, pages 58–63, Leuven, Belgium, June 3–5 1998.

[PB99] K. Purna and D. Bhatia. Temporal Partitioning and Scheduling Dataflow Graphs for Reconfigurable Computers. *IEEE Transactions on Computers*, 48(6):556–564, June 1999.

[PCI95] PCI Special Interest Group, Portland, Orlando. *PCI Local Bus Specification, Revision 2.1*, June 1 1995.

[Phi] Philips. Philips Semiconductor. http://www.semiconductors.com/catalog/.

[PJ99] M. Page-Jones. *Fundamentals of Object-oriented Design in UML*. Addison-Wesley, 1999.

[PL91] I. Page and W. Luk. Compiling Occam into FPGAs. In W. Moore and W. Luk, editors, *FPGAs*, pages 271–283. Abingdon EE&CS Books, England, 1991.

[Pla00] M. Platzner. Reconfigurable Accelerators for Combinatorial Problems. *IEEE Computer*, 33(4):58–60, April 2000.

[PRSV98] R. Passerone, J. Rowson, and A. Sangiovanni-Vincentelli. Automatic Synthesis of Interfaces between Incompatible Protocols. In *Proceedings of the 35th annual conference on Design automation conference*, pages 15–19. San Francisco, CA USA, June 1998.

[Pto] Ptolemy. Heterogeneous Modeling and Design. http://ptolemy.eecs.berkeley.edu.

[Qui] Quickturn. Quickturn: In Circuit Emulation. http://www.quickturn.com/tech/emulation.htm.

[Rab00] J. Rabaey. Silicon Platforms for the Next Generation Wireless Systems – What Role Does Reconfigurable Hardware Play? In *10th International Workshop on Field Programmable Logic and Applications, FPL 2000*, pages 277–285, Villach, Austria, August 28–30 2000.

[Ris98] Linda Rising. *The Patterns Handbook: Techniques, Strategies, and Applications*. Cambridge University Press, 1998.

[RLG+98] C. Rupp, M. Landguth, T. Garverick, E. Gomersall, H. Holt, J. Arnold, and M. Gokhale. The NAPA Adaptive Processing Architecture. In *IEEE Symposium on FPGAs for Custom Computing Machines*, pages 28–37, Napa Valley, California, April 15–17 1998.

[Röw00] T. Röwer. *Programmable Intellectual Property Modules for System Design by Reuse*. PhD thesis, ETH Zurich, Switzerland, Series In Microelectronics, Volume 105, Hartung-Gorre, 2000.

[RST+00] T. Röwer, M. Stadler, M. Thalmann, H. Kaeslin, N. Felber, and W. Fichtner. A New Paradigm for Very Flexible SONET/SDH IP-Modules. In *Proceedings of the IEEE Custom Integrated Circuits Conference 2000*, pages 533–536, Orlando, Florida, USA, May 2000.

[RSV97] J. Rowson and A. Sangiovanni-Vincentelli. Interface-based Design. In *34th Design Automation Conference*, pages 178–183, June 9–13 1997.

[SB94] S. Singh and P. Bellec. Virtual hardware for graphics applications using FPGAs. In *IEEE Workshop on FPGAs for Custom Computing Machines (FCCM)*, pages 49–58, April 10–13 1994.

[SB95] M. Srivastava and R. Brodersen. SIERA: A unified framework for rapid-prototyping of system-level hardware and software. *IEEE Transactions on Computer-Aided Design*, 14(6):676–693, June 1995.

[Sha] Mark Shand. PCI Pamette V1. COMPAQ/DEC, http://www.research.digital.com/SRC/pamette/.

[SJV95] B. Schoner, C. Jones, and J. Villasenor. Issues in Wireless Video Coding using Run-Time-reconfigurable FPGAs. In *IEEE Symposium on FPGAs for Custom Computing Machines*, pages 85–89, 1995.

[SSH+99] E. Sanchez, M. Sipper, J. Haenni, J. Beuchat, A. Stauffer, and A. Perez-Uribe. Static and Dynamic Configurable Systems. *IEEE Transactions on Computers*, 48(6):556–564, June 1999.

[Str00] Karsten Strehl. *Symbolic Methods Applied to Formal Verification and Synthesizing Embedded Systems Design*. PhD thesis, Swiss Federal Institute of Technology (ETH) Zurich, February 17 2000. Diss. ETH No. 13572. Supervised by L. Thiele and R. Ernst. *Published as:* TIK Publications Series No. 36, Shaker Verlag, 2000.

[Sun] Sundance. PCI based Products. http://www.sundance.com/html/pci-pmc_modules.htm.

[SV95] Ramesh Sivakolundu and Sunder Velamuri. *MICROWIRE/PLUS Serial Interface for COP800 Family, AN-579*. National Semiconductor Corp., Santa Clara, CA, USA, http://www.nsc.com, November 1995.

[SV98] S. Scalera and J. Vazquez. The design and implementation of a context switching FPGA. In *IEEE Symposium on FPGAs for Custom Computing Machines*, pages 78–85, Napa Valley, California, April 15–17 1998.

[SYN] SYNOPSYS. http://www.synopsys.com.

[Tan89] A. Tannenbaum. *Computer Networks*. Prentice Hall, 1989.

[TBT97] J. Teich, T. Blickle, and L. Thiele. An Evolutionary Approach to System-Level Synthesis. In *Proceedings of Codes/CASHE'97 – the 5th International Workshop on Hardware/Software Codesign*, pages 167–171, Braunschweig, Germany, March 1997.

[Tch98] Seang Tchang. Graphical User Interface in JAVA for Embedded Systems. Master's thesis, ETH Zurich, 1998. Supervisor: M. Eisenring.

[TCJW97] S. Trimberger, D. Carberry, A. Johnson, and J. Wong. A time-multiplexed fpga. In *IEEE Symposium on FPGAs for Custom Computing Machines*, pages 22–28, Napa Valley, California, April 16–18 1997.

[Tob99] C. Tobescu. Entwurfsautomatisierung für Multi-FPGA Board. Diploma thesis, Computer Engineering and Networks Laboratory, 1999. Supervisor: M. Eisenring.

[TOJH96] K. Tammemäe, M. O'Nils, A. Jantsch, and A. Hemani. Akka: A tool-kit for cosynthesis and prototyping. In *IEE Digest No.96/036 of Colloquium on Hardware-software Cosynthesis for Reconfigurable Systems*, pages 8/1–8/8, February 22 1996.

[Tri] Triscend. Configurable System-on-Chip Technology. http://www.triscend.com/products/Index.html.

[Tsc99] P. Tscharner. Report Generator für Sourcecode. Term thesis, Computer Engineering and Networks Laboratory, 1999. Supervisor: M. Eisenring.

[TSZ$^+$99] L. Thiele, Karsten Strehl, Dirk Ziegenbein, Rolf Ernst, and Jürgen Teich. *FunState* – an internal design representation forcodesign. In *Proceedings of the IEEE/ACM International Conference on Computer-Aided Design (ICCAD-99)*, pages 558–565, San Jose, California, November 7–11 1999.

[VG98] F. Vahid and T. Givargis. Incorporating Cores into System-Level Specification. In *11th International Symposium on System Synthesis ISSS'98, Hsinchu, Taiwan*, pages 43–48, December 1998.

[VLM96a] S. Vercauteren, B. Lin, and H. De Man. Constructing Application-Specific Heterogeneous Embedded Architectures from Custom HW/SW Applications. In *33rd Design Automation Conference*, pages 521–526, June 1996.

[VLM96b] S. Vercauteren, B. Lin, and H. De Man. Embedded Architecture Co-Synthesis and System Integration. In *Fourth International Workshop on Hardware/Software Co-Design*, pages 2–9, March 1996.

[VRBM96] D. Verkest, K. Van Rompaey, I. Bolsens, and H. De Man. CoWare – A Design Environment for Heterogeneous Hardware/Software Systems. *Design Automation for Embedded Systems, Kluwer Academic Publishers*, 1:357–386, October 1996.

[VSI] VSI (Virtual Socket Interface) Alliance. http://www.vsi.org/.

[VT97] F. Vahid and L. Tauro. An Object-Oriented Communication Library for Hardware-Software CoDesign. In *5th International Workshop on Hardware/Software Codesign*, pages 81–86, Braunschweig, Germany, March 1997.

[WC96] R. Wittig and P. Chow. OneChip: An FPGA Processor with Reconfigurable Logic. In *IEEE Symposium on FPGAs for Custom Computing Machines*, pages 126–135, Napa Valley, California, April 1996.

[Web00] Wolf-Dietrich Weber. Enabling Reuse via an IP Core-centric Communication Protocol: Open Core Protocol. In *IP2000, Santa Clara, CA, USA*, March 2000.

[WH96] M. Wirthlin and B. Hutchings. Sequencing Run-Time Reconfigured Hardware with Software. In *Proceedings of the 1996 ACM fourth international symposium on Field-programmable gate arrays*, pages 122–128, Monterey, CA, USA, February 11–13 1996.

[Wol97] W. Wolf. An Architectural Co-Synthesis Algorithm for Distributed, Embedded Computing Systems. *IEEE Transactions on VLSI*, 5(2):218–229, June 1997.

[XILa] XILINX. http://www.xilinx.com.

[XILb] XILINX. Press Release: New Generation of Integrated Circuits. http://www.xilinx.com/prs_rls/ibmpartner.htm.

[YW95] T. Yen and W. Wolf. Communication Synthesis for Distributed Embedded Systems. In *Proceedings of the IEEE/ACM International Conference on Computer-Aided Design (ICCAD-95)*, pages 288–294, November 5–9 1995.

[Zep95] P. Zepter. *Programmgestützter Entwurf integrierter Schaltungen für die digitale Nachrichtenübertragung aus Datenflussbeschreibungen*. PhD thesis, Lehrstuhl für Integrierte Systeme der Signalverarbeitung, TH Aachen, Germany, 1995.

[ZG97] Y. Zorian and R. Gupta. Introduction to core-based design. *IEEE Design and Test of Computers*, 14(4):15–25, October 1997.

[ZT99] E. Zitzler and L. Thiele. Multiobjective Evolutionary Algorithms: A Comparative Case Study and the Strength Pareto Approach. *IEEE Transactions on Evolutionary Computation*, 3(4):257–271, November 1999.